Troubleshooting and Repairing Diesel Engines

Troubleshooting and Repairing Diesel Engines

Fifth Edition

Paul Dempsey

Mc Graw Hill Education

New York   Chicago   San Francisco   Athens   London   Madrid
Mexico City   Milan   New Delhi   Singapore   Sydney   Toronto

Library of Congress Control Number: 2017961131

Troubleshooting and Repairing Diesel Engines, Fifth Edition

8 9 LCR 24

ISBN 978-1-260-11643-4
MHID 1-260-11643-3

The pages within this book were printed on acid-free paper.

Sponsoring Editor
Lauren Poplawski

Editorial Supervisor
Donna M. Martone

Acquisitions Coordinator
Elizabeth Houde

Project Manager
Radhika Jolly, Cenveo® Publisher Services

Copy Editor
Cenveo Publisher Services

Proofreader
Cenveo Publisher Services

Indexer
Robert Swanson

Production Supervisor
Pamela A. Pelton

Composition
Cenveo Publisher Services

Art Director, Cover
Jeff Weeks

About the author

Paul Dempsey is a master mechanic and the author of more than 20 technical books including *Small Gas Engine Repair* (now in its Second Edition), and *How to Repair Brigss & Stratton Engines* (now in its Fourth Edition), both available from McGraw-Hill. He has also written more than 100 magazine and journal articles on topics ranging from teaching techniques to maintenance management to petroleum-related subjects.

Contents

Foreword

In a world of throwaway consumer products, diesel engines are an exception. Industrial engines, those built by established manufacturers such as Caterpillar, Cummins, Deutz and Daimler run for decades with only occasional repairs. Several of these have been used in American pickup trucks, although car makers prefer in-house power. The Ford-designed 6.7L Power Stroke follows industrial practice and, as a result, is in process of receiving a B10 rating, which means that 90% of them should run for 500,000 miles without having the cylinder heads or oil pan disturbed. Smaller engines intended for commercial use have something of the same durability.

The subjects covered include:
- Diesel operation (what distinguishes diesel engines from spark-ignition engines)
- How to install stationary and marine engines
- Basic troubleshooting
- Cylinder head and engine rebuilding
- Mechanical fuel systems
- Electricity for those who are new to the subject
- Electronic fuel systems
- Turbochargers and associated air systems
- Starting and generating systems
- Air and liquid cooling systems
- Emission controls

This book is intended to supplement factory shop manuals, most of which are written cook-book style with little or nothing by way of explanation. Cook books are okay, if the only engine you will ever work on is the one you have a manual for. My aim in writing was to combine "how-to" instructions with theory. Understanding is the best, most essential tool a mechanic can have.

The more you know the easier the work becomes and the less money you waste on throwing parts at the problem. And should the job appear too demanding, an understanding of what's involved and a familiarity of the vocabulary puts shop mechanics on notice that they are dealing with a knowledgeable customer who will not be taken advantage of.

That said, diesel engines are simple mechanical devices, differing from gasoline engines only in the precision of their parts. Most repairs can be accomplished with no more than a good set of hand tools. Things get complicated when dealing with fuel systems. Special tools are needed together with an appreciation of how these systems work. You must also be aware of the hazards presented by high-pressure fuel and the lethal voltages that are sometimes present. But the rewards of working on these beautiful engines are real. Not only will you save money—shop labor charges can top $150 an hour—you will have the satisfaction that comes with accomplishment.

Paul Dempsey
Houston, TX

1
CHAPTER

Rudolf Diesel

Rudolf Diesel was born of German parentage in Paris in 1858. His father was a self-employed leather worker who, by all accounts, managed to provide only a meager income for his wife and three children. Their stay in the City of Light was punctuated by frequent moves from one shabby flat to another. Upon the outbreak of the Franco-Prussian War in 1870, the family became political undesirables and was forced to emigrate to England. Work was almost impossible to find, and in desperation, Rudolf's parents sent the boy to Augsburg to live with an uncle. There he was enrolled in school.

Diesel's natural bent was for mathematics and mechanics. He graduated as the head of his class, and on the basis of his teachers' recommendations and a personal interview by the Bavarian director of education, he received a scholarship to the prestigious Polytechnikum in Munich.

His professor of theoretical engineering was the renowned Carl von Linde, who invented the ammonia refrigeration machine and devised the first practical method of liquefying air. Linde was an authority on thermodynamics and high-compression phenomena. During one of his lectures he remarked that the steam engine had a thermal efficiency of 6–10%; that is, one-tenth or less of the heat energy of its fuel was used to turn the crankshaft, and the rest was wasted. Diesel made special note of this fact. In 1879 he asked himself whether heat could not be directly converted into mechanical energy instead of first passing through a working fluid such as steam.

On the final examination at the Polytechnikum, Diesel achieved the highest honors yet attained at the school. Professor Linde arranged a position for the young diploma engineer in Paris, where, in few months, he was promoted to general manager of the city's first ice-making plant. Soon he took charge of distribution of Linde machines over southern Europe.

By the time he was thirty, Diesel had married, fathered three children, and was recognized throughout the European scientific community as one of the most gifted engineers of the period. He presented a paper at the Universal Exposition held in Paris in 1889—the only German so honored. When he received the first of several citations of merit from German universities, he announced wryly in his acceptance speech: "I am an iceman. . ."

The basis of his acclaim was his preeminence in the new technology of refrigeration, his several patents, and a certain indefinable air about the young man that marked him as extraordinary. He had a shy, self-deprecating humor and an absolute passion for factuality. Diesel could be abrupt when faced with incompetence and was described by relatives as "proud." At the same time, he was sympathetic to his workers and made friends among them. It was not unusual for Diesel to wear the blue cotton twill that was the symbol of manual labor in the machine trades.

He had been granted several patents for a method of producing clear ice, which, because it looked like natural ice, was much in demand by the upper classes. Professor Linde did not approve of such frivolity, and Diesel turned to more serious concerns. He spent several years in Paris, working on an ammonia engine, but was defeated by the corrosive nature of this gas at pressure and high temperatures.

The theoretical basis of this research was a paper published by N.L.S. Carnot in 1824. Carnot set himself to the problem of determining how much work could be accomplished by a heat engine employing repeatable cycles. He conceived the engine drawn in Figure 1-1. Body 1 is a boiler or other heat source that raises the temperature of the working fluid A. The piston is at position C in the drawing. As the air in the cylinder is heated, it expands in correspondence to Boyle's law. If we assume a frictionless engine, the temperature of the air does not rise. Instead, expansion will take place, driving the piston to D. Then A is removed, and momentum takes the piston to E. The air column is now placed into contact with 2, which can be a radiator or cooling tank. At this point the temperature of the air falls until it exactly matches cold surface 2. The piston falls because cooled air occupies less volume than heated air. Note, however, that the temperature of the air does not change. The increase in compression as the piston falls restores heat to the air to hold its temperature constant. At B cold body 2 is removed,

1-1 Carnot cycle.

and the piston falls to A. During this phase the air gains temperature until it is equal to the heat source 1. The piston climbs back into the cylinder.

The temperature of the air, and consequently the pressure, is higher during expansion than during compression. Because the pressure is greater during expansion, the power produced by the expansion is greater than that consumed by the compression. The net result is a power output that is available for driving other machinery.

Of course this is an "ideal" cycle. It does not take into account mechanical friction nor transfer of heat from the air to the piston and cylinder walls. The difference in heat between 1 and 2 is sufficient to establish a gradient and drive the engine. It would be completely efficient.

In 1892 and 1893 Diesel obtained patent specifications from the German government covering his concept for a new type of *Verbrennungskraftmaschinen*, or heat engine. The next step was to build one. At the insistence of his wife, he published his ideas in a pamphlet and was able to interest the leading Augsburg engine builder in the idea. A few weeks later the giant Krupp concern opened negotiations. He signed another contract with the Sulzer Brothers of Switzerland.

The engine envisioned in the pamphlet and protected by the patent specifications had the following characteristics:

- *Compression of air prior to fuel delivery.* The compression was to be adiabatic; that is, no heat would be lost to the piston crown or cylinder head during this process.
- *Metered delivery of fuel, so compression pressures would not be raised by combustion temperatures.* The engine would operate on a *constant-pressure* cycle; expanding gases would keep precisely in step with the falling piston. This is a salient characteristic of Carnot's ideal gas cycle, and stands in contrast to the Otto cycle, in which combustion pressures rise so quickly upon spark ignition that we describe it as a *constant-volume* engine.
- *Adiabatic expansion.*
- *Instantaneous exhaust at constant volume.*

It is obvious that Diesel did not expect a working engine to attain these specifications. Adiabatic compression and exhaust phases are, by definition, impossible unless the engine metal is at combustion temperature. Likewise, fuel metering cannot be so precise as to limit combustion pressures to compression levels. Nor can a cylinder be vented instantaneously. But these specifications are significant in that they demonstrate an approach to invention. The rationale of the diesel engine was to save fuel by as close an approximation to the Carnot cycle as materials would allow. The steam, or *Rankine cycle*, engine was abysmal in this regard; and the *Otto* four-stroke spark or hot-tube ignition engine was only marginally better.

This approach, from the mathematically ideal to materially practical, is exactly the reverse of the one favored by inventors of the Edison, Westinghouse, and Kettering school. When Diesel visited America in 1912, Thomas A. Edison explained to the young inventor that these men worked *inductively*, from the existing technology, and not *deductively*, from some ideal or model. Diesel felt that such procedure was at best haphazard, even though the results of Edison and other inventors of the inductive school were obviously among the most important.

The first Diesel engine was a single-cylinder four-cycle design, operated by gasoline vapor. The vapor was sprayed into the cylinder near top dead center by means of an air compressor. The engine was in operation in July of 1893. However, it was discovered that a misreading of the blueprints had caused an increase in the size of the chamber. This was corrected with a new piston, and the engine was connected to a pressure gauge. The gauge showed approximately 80 atmospheres before it shattered, spraying the room with brass and glass fragments. The best output of what Diesel called his "black mistress" was slightly more than 2 hp—not enough power to overcome friction and compression losses. Consequently, the engine was redesigned.

The second model was tested at the end of 1894. It featured a variable-displacement fuel pump to match engine speed with load. In February of the next year, the mechanic Linder noted something remarkable. The engine had been sputtering along, driven by a belt from the shop power plant, but Linder noticed that the driving side of the belt was slack, indicating that the engine was putting power into the system. For the first time, a Diesel engine ran on its own.

Careful tests—and Diesel was nothing if not careful and methodical—showed that combustion was irregular. The next few months were devoted to redesigning the nozzle and delivery system. This did not help, and in what might have been a fit of desperation, Diesel called upon Robert Bosch for an ignition magneto. Bosch personally fitted one of his low-tension devices to the engine, but it had little effect on the combustion problem. Progress came about by varying the amount of air injected with the fuel, which, at this time, was limited to kerosene or gasoline.

A third engine was built with a smaller stroke/bore ratio and fitted with two injectors. One delivered liquid fuel, the other a mixture of fuel and air. It was quite successful, producing 25 hp at 200 rpm. Further modifications of the injector, piston, and lubrication system ensued, and the engine was deemed ready for series production at the end of 1896.

Diesel then turned his attention to his family, music, and photography. Money began to pour in from the patent licensees and newly organized consortiums wanting to build engines in France, England, and Russia. The American brewer Adolphus Busch purchased the first commercial engine, similar to the one on display at the Budweiser plant in St. Louis today. He acquired the American patent rights for one million marks, which at the current exchange rate amounted to a quarter of a million dollars—more than Diesel had hoped for.

The next stage of development centered around various fuels. Diesel was already an expert on petroleum, having researched the subject thoroughly in Paris in an attempt to refine it by extreme cold. It soon became apparent that the engine could be adapted to run on almost any hydrocarbon from gasoline to peanut oil. Scottish and French engines routinely ran on shale oil, while those sold to the Nobel combine in Russia operated on refinery tailings. In a search for the ultimate fuel, Diesel attempted to utilize coal dust. As dangerous as this fuel is in storage, he was able to use it in a test engine.

These experiments were cut short by production problems. Not all the licensees had the same success with the engine. In at least one instance, a whole production run had to be recalled. The difficulty was further complicated by a shortage of trained technicians. A small malfunction could keep the engine idle for weeks, until the customer lost patience and sent it back to the factory. With these embarrassments came the question of whether the engine had been oversold. Some believed that it needed much more development before being put on the market. Diesel was confident that his creation was practical—if built and serviced to specifications. But he encouraged future development by inserting a clause in the contracts that called for pooled research—the licensees were to share the results of their research on Diesel engines.

Diesel's success was marred in two ways. For one, he suffered exhausting patent suits. The Diesel engine was not the first to employ the principle of compression ignition; Akroyd Stuart had patented a superficially similar design in 1890. Also, Diesel had a weakness for speculative investments. This weakness, along with a tendency to maintain a high level of personal consumption, cost Rudolf Diesel millions. His American biographers, W. Robert Nitske and Charles Morrow Wilson, estimate that the mansion in Munich cost a million marks to construct at the turn of the century.

The inventor eventually found himself in the uncomfortable position of living on his capital. His problem was analogous to that of an author who is praised by the critics but who cannot seem to sell his books. Diesel engines were making headway in stationary and marine applications, but they were expensive to build and required special service techniques. True mass production was out of the question. At the same time, the inventor had become an international celebrity, acclaimed on three continents.

Diesel returned to work. After mulling a series of projects, some of them decidedly futuristic, he settled on an automobile engine. Two such engines were built. The smaller, 5-hp model was put into production, but sales were disappointing. The engine is, by nature of its compression ratio, heavy and, in the smaller sizes, difficult to start. (The latter phenomenon is due to the unfavorable surface/volume ratio of the chamber as piston size is reduced. Heat generated by compression tends to bleed off into the surrounding metal.) A further complication was the need for compressed air to deliver the fuel into the chamber. Add to these problems associated with precision machine work, and the diesel auto engine seemed impractical. While diesel trucks appeared early, it would be 1936 before Mercedes-Benz produced the first commercially successful diesel passenger car.

Diesel worked for several months on a locomotive engine built by the Sulzer Brothers in Switzerland. First tests were disappointing, but by 1914 the Prussian and Saxon State Railways had a diesel in regular service. Of course, nearly all of the world's locomotives are diesel-powered today.

Maritime applications came as early as 1902. Nobel converted some of his tanker fleet to diesel power, and by 1905 the French navy was relying on these engines for their submarines. Seven years later, almost 400 boats and ships were propelled solely or in part by compression engines. The chief attraction was the space saved, which increased the cargo capacity or range.

In his frequent lectures Diesel summed up the advantages of his invention. The first was efficiency, which was beneficial to the owner and, by extension, to all of society. In immediate terms, efficiency meant cost savings. In the long run, it meant conserving world resources. Another advantage was that compression engines could be built on any scale from the fractional horsepower to the 2400-hp Italian Tosi of 1912. Compared to steam engines, the diesel was compact and clean. Rudolf Diesel was very much concerned with the question of air pollution, and mentioned it often.

But the quintessential characteristic, and the one that might explain his devotion to his "black mistress," was her quality. Diesel admitted that the engines were expensive, but his goal was to build the best, not the cheapest.

During this period Diesel turned his attention to what his contemporaries called "the social question." He had been poor and had seen the effects of industrialization firsthand in France, England, and Germany. Obviously machines were not freeing men, or at least not the masses of men and women who had to regulate their lives by the factory system. This paradox of greater output of goods and intensified physical and spiritual poverty had been seized on by Karl Marx as the key "contradiction" of the capitalistic system. Diesel instinctively distrusted Marx because he distrusted the violence that was implicit in "scientific socialism." Nor could he take seriously a theory of history whose exponent claimed it was based on absolute principles of mathematical integrity.

He published his thoughts on the matter under the title *Solidarismus* in 1903. The basic concept was that nations were more alike than different. The divisions that characterize modern society are artificial to the extent that they do not have an economic rationale. To find solidarity, the mass of humanity must become part owners in the sources of production. His formula was for every worker to save a penny a day. Eventually these pennies would add up to shares or part shares in business enterprises: Redistributed, wealth and, more important, the sense of controlling one's destiny would be achieved without violence or rancor through the effects of the accumulated capital of the workers.

Diesel wrote another book that was better received. Entitled *Die Enstehung des Dieselmotors*, it recounted the history of his invention and was published in the last year of his life.

For years Diesel had suffered migraine headaches, and in his last decade, he developed gout, which at the end forced him to wear a special oversized slipper. Combined with this was a feeling of fatigue, a sense that his work was both done and undone, and that there was no one to continue. Neither of his two sons showed any interest in the engine, and he himself seemed to have lost the iron concentration of earlier years when he had thought nothing of a 20-hour workday. It is probable that technicians in the various plants knew more about the current state of diesel development than he did.

And the bills mounted. A consultant's position, one that he would have coveted in his youth, could only postpone the inevitable; a certain level of indebtedness makes a salary superfluous. Whether he was serious in his acceptance of the English-offered consultant position is unknown. He left his wife in Frankfort in apparent good spirits and gave her a present. It was an overnight valise, and she was instructed not to open it for a week. When she did, she found it contained 20,000 marks. This was, it is believed, the last of his liquid reserves. At Antwerp he boarded the ferry to Warwick in the company of three friends. They had a convivial supper on board and retired to their staterooms. The next morning Rudolf Diesel could not be found. One of the crew discovered his coat, neatly folded under a deck rail. A few days later a pilot boat sighted a body floating in the channel, removed a corn purse and spectacle case from the pockets, and set the corpse adrift. The action was not unusual or callous; seamen had, and still do have, a horror of retrieving bodies from the sea. These items were considered by the family to be positive identification. They accepted the death as suicide, although the English newspapers suggested foul play at the hands of foreign agents who did not want Diesel's engines in British submarines.

2
CHAPTER

Diesel basics

At first glance, a diesel engine looks like a heavy-duty gasoline engine, minus spark plugs and ignition wiring (Figure 2-1). Some manufacturers build compression ignition (CI) and spark ignition (SI) versions of the same engine. Caterpillar G3500 and G3600 SI natural-gas fueled engines are built on diesel frames and use the same blocks, crankshafts, heads, liners, and connecting rods.

But there are important differences between CI and SI engines that cut deeper than the mode of igniting the fuel.

Compression ratio

When air is compressed, collisions between molecules produce heat that ignites the diesel fuel. The compression ratio (c/r) is the measure of how much the air is compressed (Figure 2-2).

Compression ratio = swept volume + clearance volume ÷ swept volume

Swept volume = the volume of the cylinder traversed by the piston in its travel from top dead center (tdc) to bottom dead center (bdc)

Clearance volume = combustion chamber volume

Figure 2-3 graphs the relationship between c/r's and thermal efficiency, which reaffirms what every mechanic knows—high c/r's are a precondition for power and fuel economy.

At the very minimum, a diesel engine needs a c/r of about 16:1 for cold starting. Friction, which increases more rapidly than the power liberated by increases in compression, sets the upper limit at about 24:1. Other inhibiting factors are the energy required for cranking and the stresses produced by high power outputs. Diesels with c/r's of 16 or 17:1 sometimes benefit from a point or two of higher compression. Starting becomes easier and less exhaust smoke is produced. An example is the

1. Cylinder head
2. Cylinder body
3. Main bearing housing
4. Exhaust valve
5. Intake valve
6. Valve spring
7. Valve rocker arm support
8. Valve rocker arm
9. Precombustion chamber
10. Decompression lever

11. Intake silencer
12. Mixing elbow
13. Camshaft
14. Camshaft gear
15. Tappet
16. Push rod
17. Piston
18. Connecting rod
19. Crankshaft
20. Crankshaft gear

21. Flywheel
22. Crankshaft V-pulley
23. Oil pan
24. Dipstick
25. Lubricating oil pump
26. Lubricating oil inlet pipe
27. Anti corrosion zinc
28. Fuel injection pump cam
29. Fuel injection pump
30. Fuel injection nozzle

31. Fuel feed pump
32. Cooling water pump
33. Thermostat
34. Starter motor
35. Damper disc
36. Input shaft
37. Output shaft
38. Forward large gear
39. Reverse large gear
40. Output shaft coupling

2-1 The Yanmar 1GM10, shown with a marine transmission, provides auxiliary power for small sailboats. The 19.4 CID unit develops 9 hp and forms the basic module for two- and three-cylinder versions.

$$\varepsilon = \frac{VC + VS}{VC} = 1 + \frac{VS}{VC} \text{ (equation 1)}$$

2-2 Compression ratio is a simple concept, but one that mathematics and pictures express better than words.

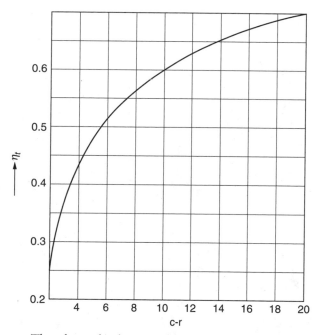

2-3 The relationship between diesel compression ratios and thermal efficiency.

Caterpillar 3208 that has a tendency to smoke and "wet stack," that is, to saturate its exhaust system with unburned fuel. These problems can be alleviated with longer connecting rods that raise the compression ratio from 16.5:1 to 18.2:1.

It should be noted that a compressor, in the form of a turbocharger or supercharger, raises the effective c/r. Consequently, these engines have c/r's of 16 or 17:1, which are just adequate for starting. Once the engine is running, the compressor provides additional compression.

Gasoline engines have lower c/r's—half or less—than CI engines. This is because the fuel detonates when exposed to the heat and pressure associated with higher c/r's. Detonation is a kind of maverick combustion that occurs after normal ignition. The unburned fraction of the charge spontaneously explodes. This sudden rise in pressure can be heard as a rattle or, depending upon the natural frequency of the connecting rods, as a series of distinct pings. Uncontrolled detonation destroys crankshaft bearings and melts piston crowns.

Induction

Most SI engines mix air and fuel in the intake manifold by way of one or more low-pressure (50-psi or so) injectors. A throttle valve regulates the amount of air admitted, which is only slightly in excess of the air needed for combustion. As the throttle opens, the injectors remain open longer to increase fuel delivery. For a gasoline engine, the optimum mixture is roughly 15 parts air to 1 part fuel. The air-fuel mixture then passes into the cylinder for compression and ignition.

In a CI engine, air undergoes compression before fuel is admitted. Injectors open late during the compression stroke as the piston approaches tdc. Compressing air, rather than a mix of air and fuel, improves the thermal efficiency of diesel engines. To understand why would require a course in thermodynamics; suffice to say that air contains more latent heat than does a mixture of air and vaporized fuel.

Forcing fuel into a column of highly compressed air requires high injection pressures. These pressures range from about 6000 psi for utility engines to as much as 30,000 psi for state-of-the-art examples.

CI engines dispense with the throttle plate—the same amount of air enters the cylinders at all engine speeds. Typically, idle-speed air consumption averages about 100 lb of air per pound of fuel; at high speed or under heavy load, the additional fuel supplied drops the ratio to about 20:1.

Without a throttle plate, diesels breathe easily at low speeds, which explains why truck drivers can idle their rigs for long periods without consuming appreciable fuel. (An SI engine requires a fuel-rich mixture at idle to generate power to overcome the throttle restriction.)

Since diesel air flow remains constant, the power output depends upon the amount of fuel delivered. As power requirements increase, the injectors deliver more fuel than can be burned with available oxygen. The exhaust turns black with partially oxidized fuel. How much smoke can be tolerated depends upon the regulatory climate, but the smoke limit always puts a ceiling on power output.

To get around this restriction, many diesels incorporate an air pump in the form of an exhaust-driven turbocharger or a mechanical supercharger. Forced induction can double power outputs without violating the smoke limit. And, as far as turbochargers are concerned, the supercharge effect is free. That is, the energy that drives the turbo would otherwise be wasted out the exhaust pipe as heat and exhaust-gas velocity.

The absence of an air restriction and an ignition system that operates as a function of engine architecture can wrest control of the engine from the operator. All that's needed is for significant amounts of crankcase oil to find its way into the combustion chambers. Oil might be drawn into the chambers past worn piston rings or from a failed turbocharger seal. Some industrial engines have an air trip on the intake manifold for this contingency, but many do not. A runaway engine generally accelerates itself to perdition because few operators have the presence of mind to engage the air trip or stuff a rag into the intake.

Ignition and combustion

SI engines are fired by an electrical spark timed to occur just before the piston reaches the top of the compression stroke. Because the full charge of fuel and air is present, combustion proceeds rapidly in the form of a controlled explosion. The rise in cylinder pressure occurs during the span of a few crankshaft degrees. Thus, the cylinder volume above the piston undergoes little change between ignition and peak pressure. Engineers, exaggerating a bit, describe SI engines as "constant volume" engines (Figure 2-4).

Compared to SI, the onset of diesel ignition is a leisurely process (Figure 2-4). Some time is required for the fuel spray to vaporize and more time is required for the spray to reach ignition temperature. Fuel continues to be injected during the delay period.

Once ignited, the accumulated fuel burns rapidly with correspondingly rapid increases in cylinder temperature and pressure. The injector continues to deliver fuel through the period of rapid combustion and into the period of controlled combustion that follows. When injection ceases, combustion enters what is known as the afterburn period.

The delay between the onset of fuel delivery and ignition (A–B in Figure 2-5) should be as brief as possible to minimize the amount of unburnt fuel accumulated in the cylinder. The greater the ignition lag, the more violent the combustion and resulting noise, vibration, and harshness (NVH).

Ignition lag is always worst upon starting cold, when engine metal acts as a heat sink. Mechanics sometimes describe the clatter, white exhaust smoke, and rough combustion that accompany cold starts as "diesel detonation," a term that is misleading because diesels do not detonate in the manner of SI engines. Combustion should smooth out after the engine warms and ignition lag diminishes. Heating the incoming air makes cold starts easier and less intrusive.

In normal operation, with ignition delay under control, cylinder pressures and temperatures rise more slowly (but to higher levels) than for SI engines. In his proposal of 1893, Rudolf Diesel went one step further and visualized constant pressure expansion—fuel input and combustion pressure would remain constant during the expansion, or

1 ──►2   Intake stroke
2 ──►3   Compression stroke
3 ──►4   Combustion
      Otto cycle: Constant volume change
      Diesel cycle: Constant pressure
            change
4 ──►5   Expansion stroke
5 ►2►1   Exhaust stroke

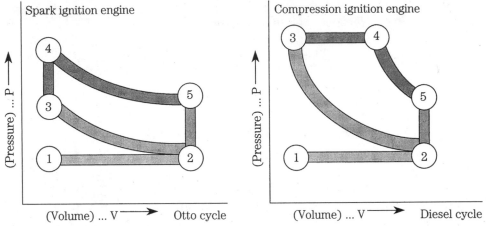

2-4 These cylinder pressure/volume diagrams distort reality somewhat, but indicate why SI engines are described as "constant volume" and CI as "constant pressure."

power, stroke. He was able to approach that goal in experimental engines, but only if rotational speeds were held low. His colleagues eventually abandoned the idea and controlled fuel input pragmatically, on the basis of power output. Even so, the pressure rise is relatively smooth and diesel engines are sometimes called "constant pressure" devices to distinguish them from "constant volume" SI engines (shown at Figure 2-4).

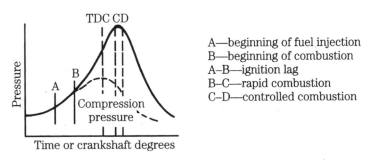

A—beginning of fuel injection
B—beginning of combustion
A–B—ignition lag
B–C—rapid combustion
C–D—controlled combustion

2-5 Diesel combustion and compression pressure rise plotted against crankshaft rotation.

Two- and four-stroke-cycle

CI and SI engines operate on similar cycles, consisting of intake, compression, expansion, and exhaust events. Four-stroke-cycle engines of either type allocate one up or down stroke of the piston for each of the four events. Two-stroke-cycle engines telescope events into two strokes of the piston, or one per crankshaft revolution. In the United States, the term *stroke* is generally dropped and we speak of two- or four-cycle engines; in other parts of the English-speaking world, the preferred nomenclature is two-stroke and four-stroke.

Four-cycle diesel engines operate as shown in Figure 2-6. Air, entering around the open intake valve, fills the cylinder as the piston falls on the intake stroke. The intake valve closes as the piston rounds bdc on the compression stroke. The piston rises, compressing and heating the air to ignition temperatures.

Injection begins near tdc on the compression stroke and continues for about 40° of crankshaft rotation. The fuel ignites, driving the piston down in the bore on the expansion, or power stroke. The exhaust valve opens and the piston rises on the exhaust stroke, purging the cylinder of spent gases. When the piston again reaches tdc, the four-stroke-cycle is complete, two crankshaft revolutions from its beginning.

Figure 2-7 illustrates the operation of Detroit Diesel two-cycle engines, which employ blower-assisted scavenging. As shown in the upper left drawing, pressurized air enters the bore through radial ports and forces the exhaust gases out through the cylinder

2-6 Four-cycle operation. Yanmar Diesel Engine Co. Ltd.

Scavenging   Compression

Power   Exhaust

2-7 Two-cycle operation. The Detroit Diesel engine depicted here
employs a Roots-type positive-displacement blower for scavenging.

without raising its pressure much above atmospheric. The exhaust valve remains open
until the ports are closed to eliminate a supercharge effect.

The exhaust valve then closes and the piston continues to rise, compressing the
air charge ahead of it. Near tdc, the injector fires, combustion begins, and cylinder
pressure peaks as the piston rounds tdc. Expanding gases drive the piston downward.
The exhaust valve opens just before the scavenge ports are uncovered to give spent
gases opportunity to blow down. These four events—intake, compression, expansion,
and exhaust—occur in two piston strokes, or one crankshaft revolution.

Not all two-cycle diesel engines have valves. Combining scavenge air with combustion air eliminates the intake valve, and a port above the air inlet port replaces the
exhaust valve. Such engines employ cross-flow or loop scavenging (Figure 2-8) to purge

2-8 Cross-flow scavenging employs a deflector on the piston crown to divert the incoming air charge up and away from the exhaust port. Loop scavenging achieves the same effect with angled inlet ports.

the upper reaches of the cylinder and to minimize the loss of scavenge air to the exhaust. In the cross-flow scheme, a deflector cast into the piston crown diverts the incoming air stream away from the open exhaust port and into the stagnant region above the piston. The angled inlet ports on loop-scavenged engines produce the same effect.

It is also possible to eliminate the external air pump by using the crankcase as part of the air inlet tract. Piston movement provides the necessary compression to pump the air, via a transfer port, into the cylinder. Not many crankcase-scavenged diesel engines are seen in this country, but the German manufacturer Fichtel & Sachs has built thousands of them.

Because two-cycle engines fire every revolution, the power output should be twice that of an equivalent four-stroke. Such is not the case, principally because of difficulties associated with scavenging. Four-cycle engines mechanically purge exhaust gases, through some 440° of crankshaft revolution. (The exhaust valve opens early during the expansion stroke and closes after the intake valve opens.) Two- cycles scavenge in a less positive manner during an abbreviated interval of about 130°. Consequently, some exhaust gas remains in the cylinder to dampen combustion.

Power and torque

Horsepower is the ability to perform work over time. In 1782, James Watt, a pioneer developer of steam engines, observed that one mine pony could lift 550 lb of coal one foot in one minute. Torque is the instantaneous twisting force applied to the crankshaft. In the English-speaking world, we usually express torque as pounds of force applied on a lever one foot long.

The two terms are related as follows:

Horsepower = torque × 2pi × rpm. Revolutions per minute is the time component.

Torque = displacement × 4pi × bmep. The latter term, brake mean effective pressure, is the average pressure applied to the piston during the expansion stroke.

High-performance diesels, such as used in automobiles, develop maximum horsepower at around 5000 rpm. Equivalent SI auto engines can turn almost twice as fast. Since rpm is part of the horsepower formula, these diesels fall short in the power department. An SI-powered car will have a higher top speed.

But, thanks to high effective brake mean pressures, diesels have the advantage of superior torque. A diesel-powered BMW or Mercedes-Benz easily out-accelerates its gasoline-powered cousins.

Fuel efficiency

High c/r's (or more exactly, large ratios of expansion) give CI engines superior thermal efficiency. Under optimum conditions, a well-designed SI engine utilizes about 30% of the heat liberated from the fuel to turn the crankshaft. The remainder goes out the exhaust and into the cooling system and lubricating oil. CI engines attain thermal efficiencies of 40% and greater. By this measure—which is increasingly critical as fears about global warming are confirmed—diesel engines are the most efficient practical form of internal combustion. (Gas turbines do better, but only at constant speeds.)

Excellent thermal efficiency, plus the volumetric efficiency afforded by an unthrottled intake manifold and the ability to recycle some exhaust heat by turbocharging, translate into fuel economy. It is not unreasonable to expect a specific fuel consumption of 0.35 lb/hp-hr from a CI engine operating near its torque peak. An SI engine can consume 0.50 lb/hp-hr under the same conditions.

The weight differential between diesel fuels (7.6 lb/US gal for No. 2D) and gasoline (about 6.1 lb/US gal) gives the diesel an even greater advantage when consumption is figured in gallons per hour or mile. CI passenger cars and trucks deliver about 30% better mileage than the same vehicles with gasoline engines.

Diesel pickups and SUVs appeal in ways other than fuel economy. Owners of these vehicles tend to become diesel enthusiasts. I'm not sure why, but it probably has something to do with the sheer mechanical presence that industrial products radiate. Earlier generations had the same sort of love affair with steam.

Weight

The Cummins ISB Dodge pickup motor weighs 962 lb and develops 260 hp for a wt/hp ratio of 3.7:1. The 500-hp Caterpillar 3406E, a standard power plant for large (Grade-8) highway trucks comes in at 5.7 lb/hp. The Lugger, a marine engine of legendary durability, weighs 9.6 lb for each of its 120 horses. By comparison, the Chevrolet small block SI V-8 has an all-up weight of about 600 lb and with a bit of tweaking develops 300 hp.

Much of the weight of diesel engines results from the need to contain combustion pressures and heat that, near tdc, peak out at around 1000 psi and 3600°F. And, as mentioned earlier, bmep or average cylinder pressures are twice those of SI engines.

There are advantages to being built like a Sumo wrestler. Crankshaft bearings stay in alignment, cylinder bores remain round, and time between overhauls can extend for tens of thousands of hours.

Durability

Industrial diesel engines come out of a conservative design tradition. High initial costs, weight, and moderate levels of performance are acceptable tradeoffs against early failure. The classic diesel is founded on heavy, fine-grained iron castings, liberally reinforced with webbing and aged prior to machining. Buttressed main-bearing caps, pressed into the block and often cross-drilled, support the crankshaft. Pistons run against replaceable liners, whose metallurgy can be precisely controlled. Some of the better engines, such as the Cummins shown in Figure 2-9, feature straight-cut timing gears, which are virtually indestructible.

Heavy truck piston rings go for a million miles between replacements. An early Caterpillar 3176 truck engine was returned to the factory for teardown after logging more than 600,000 miles. Main and connecting-rod bearings had been replaced (at 450,000 and 225,000 miles, respectively) and were not available for examination. The parts were said to be in good condition.

The crankshaft remained within tolerance, as did the rocker arms, camshaft journals, and lower block casting. Valves showed normal wear, but were judged reusable. Connecting rods could have gone another 400,000 miles and pistons for 200,000 miles. The original honing marks were still visible on the cylinder liners.

But Caterpillar was not satisfied, and made a series of major revisions to the 3176, including redesigned pistons, rings, connecting rods, head gasket, rocker shafts, injectors, and water pump. Crankshaft rigidity has been improved, and tooling developed to give the cylinder liners an even more durable finish.

2-9 The Cummins ISB employs straight-cut timing gears that, while noisy, are practically indestructible. Gilmer-type toothed timing belts, typical of passenger-car diesels, need replacement at 60,000 miles or less.

Durability is not a Caterpillar exclusive—according to the EPA, heavy-truck engines have an average life cycle of 714,000 miles. Not a few Mercedes passenger cars have passed the three-quarter-million mile mark with only minor repairs.

This is not to say that diesels are zero-defect products. Industrial engines are less than perfect, and when mated with digital technology the problems multiply. Many of the worst offenders are clones, that is, diesels derived from existing SI engines. No one who was around at the time can forget the 1978 Oldsmobile Delta 88 Royale that sheared head bolts, crankshafts, and almost everything in between. Another clone that got off to a bad start was the Volkswagen. Like the Olds, it had problems with fasteners and soft crankshafts. But these difficulties were overcome. Today the VW TDi is the most popular diesel passenger-car engine in Europe accounted for 40% of Volkswagen's production.

Conventional fuels

Diesel fuel is a middle distillate, slightly heavier than kerosene or jet fuel. Composition varies with the source crude, the refining processes used, the additive mix, and the regulatory climate. ASTM (American Society for Testing Materials) norms for Nos. 1-D and 2-D fuels in the United States are shown in Table 2-1.

Table 2-2 lists characteristics the EPA considers typical for Nos. 1-D and 2-D ULSD sold outside of California, which has its own, more rigorous rules. Note that EPA regulations apply only to sulfur content and to cetane number/aromatic content.

Table 2-1. **ASTM diesel fuel grades**

Grade	Characteristics	Sulfur content
No. 1-D S15 ULSD (ultra-low sulfur diesel)	ULSD is mandatory for use on all 2007-model and later road vehicles. Because of its high volatility No. 1-D ULSD is sometimes substituted for No. 2-D ULSD in cold climates.	15 ppm
No. 1-D S500	Obsolete. Damages emission control equipment on 2007 and later model vehicle engines	500 ppm
No. 2-D S15 ULSD	The standard fuel for vehicles and other small, high-speed engines. Produces slightly more power than No. 1-D ULSD	15 ppm
No. 2-D S500	Obsolete. Damages emission control equipment on 2007 and later model vehicle engines.	500 ppm

Other fuel qualities, such as lubricity, filterability, and viscosity, are left to the discretion of the refiner. As a general rule, large truck stops provide the best, most consistent fuel.

- Cetane number (CN) and aromatic content refer to the ignition quality of the fuel. U.S. regulations permit 40 CN fuel if the aromatic content does not exceed 35%. In Europe diesel fuel must have a CN of at least 51. Aromatic content expresses the ignition quality of the fuel. High-octane fuels, such as aviation gasoline, have low CNs and barely support diesel combustion. Conversely, ether and amyl nitrate, which detonate violently in SI engines, are widely used as diesel starting fluids.
- API (American Petroleum Institute) gravity is an index of fuel density and, by extension, its caloric value. Heavier fuels produce more energy per injected volume.
- Viscosity also affects performance. Less viscous fuels atomize better and produce less exhaust smoke. But extremely light fuel upsets calibration by leaking past pump plungers. Thick, highly viscous fuels increase delivery pressures and pumping loads.
- Flash point, or the temperature at which the fuel releases ignitable vapors, is a safety consideration.

Table 2-2. **ULSD fuel characteristics**

	No. 1D	No. 2D
Cetane number	40–54	40–50
Gravity, °API	40–44	32–37
Sulfur, ppm	7–15	7–15
Min. aromatics, %	8	27
Min. flashpoint, °F	120	130
Viscosity, centistokes	1.6–2.0	2.0–3.2

3
CHAPTER

Engine installation

This chapter describes power requirements, mounting provisions, and alignment procedures for installing diesel engines in motor vehicles, stationary applications, and small boats. What I have tried to do here is to provide information that does not have wide currency, but is so critical that it makes or breaks the installation. Vendor catalogs serve for other aspects of the job, such as radiator/keel cooler sizing, selection of anti-vibration mounts, and sound-proofing techniques.

Trucks and other motor vehicles

Normally, installation is a bolt-on proposition, but things become complex when engines or transmissions are not as originally supplied.

Power requirements

Operators often judge a truck's power, or lack of it, by how fast the truck runs. In other words, operators look at maximum rated horsepower available at full governed rpm. But expected road speeds may be unrealistic. For example, numerically low axle ratios can, up to a point, increase top speed, but at the cost of reduced acceleration and less startability, a term that is defined below. Other factors that influence top speed are loaded weight, road conditions, wind resistance (which can double when loads are carried outside of the vehicle bodywork), and altitude. Naturally aspirated engines lose about 3% of their rated power per 1000 ft of altitude above sea level.

The desired cruising speed should be 10–20% below rated horsepower rpm, to provide a reserve of power for hill climbing and passing. When fuel economy is a primary consideration, the cruising speed can be set even lower. The power required at cruising speed is the engine's net horsepower.

Other factors to consider are the ability of the vehicle to cope with grades. Startability is expressed as the percentage grade the vehicle can climb from a dead stop. A fully loaded general-purpose truck should be able to get moving up a 10% grade in low gear. Off-road vehicles should be able to negotiate 20% grades, with little or no clutch slippage. Startability is a function of the lowest gear ratio and the torque available at 800–1000 rpm.

Gradeability is the percentage grade a truck can climb from a running start while holding a steady speed. No vehicle claiming to be self-propelled should have a gradeability of less than 6%. Gradeability depends upon maximum torque the engine is capable of multiplied by intermediate gearing.

Caterpillar and other engine manufacturers can provide assistance for sizing the engine to the particular application. But it's useful to have some understanding of how power requirements are calculated.

The power needed to propel a vehicle is the sum of driveline losses, air resistance, rolling resistance, and grade resistance.

$$\text{driveline losses} = 1 - \text{driveline eff.} \times \text{hp}_{air} + \text{hp}_{roll} + \text{hp}_{grade}$$

driveline eff. = overall efficiency of the driveline, calculated on the assumption that each driven element—main transmission, auxiliary transmission, and rear axle—imposes an efficiency penalty of 4%. Thus, a truck with a single transmission and one driven rear axle would have an overall driveline efficiency of 92% ($0.96 \times 0.96 = 92\%$).

$$\text{hp}_{air} = \text{air resistance hp} = (\text{mph}^3 \div 375) \times 0.00172 \times \text{modifier} \times \text{frontal area}$$

Without some sort of provision to smooth airflow, the truck has a modifier of 1.0. If an aerodynamic device is fitted, the modifier is 0.60. For purposes of our calculation, frontal area = width in feet × (height in feet − 0.75 ft).

$$\text{hp}_{roll} = \text{rolling resistance hp} = \text{GVW} \times \text{mph} \times \text{Crr}$$

where GVW represents the gross vehicle weight in pounds, and Crr represents the rolling resistance. This latter figure depends upon tire type—on smooth concrete, bias-ply tires have a Crr of 17 lb/ton and radial tires 11 lb/ton. Low-profile tires do even better.

$$\text{hp}_{grade} = \text{grade hp} = (\text{grade percentage} \times \text{GVW} \times \text{mph}) \div 37{,}500$$

Motor mounts

In most instances, the technician merely bolts the engine to a factory-designed mounting system. But there are times when engine-mounting provisions cannot be taken for granted.

Vehicle engines traditionally use a three-point mounting system, with a single point forward around which the unit can pivot, and with two points at the flywheel housing or transmission. For some engines the forward mount takes the form of an extension, or trunnion, at the crankshaft centerline. A sleeve locates the trunnion laterally, while permitting the engine to rotate. In order to simplify mounting and give more control over resiliency, other engines employ a rigid bracket bolted to the timing cover and extending out either side to rubber mounts on the frame.

Rear mounts normally bolt to the flywheel cover and function to locate the engine fore and aft, while transmitting the torque reaction to the vehicle frame. In order to control vibration, mount stiffness must be on the order of one-tenth of frame stiffness.

M_X = Static Bending Moment at Flywheel Housing/Transmission
L_1 = Transmission CG Location
W_T = Weight of Transmission
F = Rear Support Lift
L_2 = Rear Support Location

Engine Front Mount

Engine Rear Mount

$$M_X = W_T L_1$$

For Static Bending Moment Calculation for Transmission Support Installation use:

$$M_X = W_T L_1 - FL_2$$

3-1 If we think of the motor mounts as springs, it is easy to see that adding a transmission mount reduces the bending forces applied to the bell housing by the weight of the transmission. However for this to happen, the transmission mount must have a lower spring rate than the rear engine mounts. A high spring rate at the transmission neutralizes the rear motor mounts so that the whole weight of the engine and transmission is shared between the front motor mounts and the rear transmission mount. Bending forces increase. And, in practice, frame members adjacent to the transmission mount bend.
Courtesy Caterpillar Inc.

On many applications the transmission cantilevers off the engine block without much additional support. The bending moment imposed by the overhung load on the flywheel housing should be calculated and compared against factory specs for the engine. Figure 3-1 illustrates the calculation for a transmission that receives some additional support at the rear with a third mount.

The third mount should have a vertical rate (lb/in. of deflection) considerably lower than the vertical rate of the rear engine mounts. A transmission mount with a higher spring rate than the engine mounts increases the bending moment. In addition, the high spring rate is almost sure to deflect the truck frame and, in the process, generate high forces on the engine/transmission package.

Off-road trucks present special problems, since engines and transmissions are subject to high gravity loads and the potential for frame distortion. Another factor that needs to be taken into consideration is that motor mounts must be able to absorb the torque reactions generated by the ultra-low gear ratios often specified for these vehicles.

Stationary engines

Power requirements

Power requirements for stationary applications can be difficult to calculate. Wherever possible, engine selection should be based upon experience and verified by tests in the field.

Caterpillar rates its industrial engines on a five-tier format based on load- factor duty cycle, annual operating hours, and expected time between overhaul. The load factor is a measure of the actual power output of the engine that, at any particular throttle setting, depends upon load. For example, an engine set to produce 300 hp will produce 50 hp under a 50-hp load, 100 hp under a 100-hp load, and so on. Fuel consumption increases with load demand. The load factor indicates how hard the engine works, and is calculated by comparing actual fuel usage with no-load usage at the throttle setting appropriate for the application.

- Industrial A—100% duty cycle under full load at rated rpm. Applications include pipeline pumping stations and mixing units for oilfield service.
- Industrial B—Maximum 80% duty cycle. Typical applications are oilfield rotary-table drives and drilling-mud, and cement pumps
- Industrial C—Maximum duty cycle 50%, with one hour at full load and speed, followed by one hour at reduced demand. Applications include off-road trucks, oilfield hoisting, and electric power generation for oil rigs.
- Industrial D—Maximum duty cycle not to exceed 10%, with up to 30 minutes of full load and power followed by 1 hour at part throttle. Used for offshore cranes and coiled-tubing drilling units where loads are cyclic.
- Industrial E—Maximum duty cycle not to exceed 5%, with no more than 15 minutes at full power, followed by one hour at reduced load. E-rated engines may need to develop full power at starting or to cope with short-term emergency demands.

All things equal, an oversized engine works at a lower load factor and should run longer between overhauls. Power in reserve also means that overloads can be accommodated without loss of rpm.

Power, the measure of the work the engine performs over time, is only part of the picture. Engines also need to develop torque, an instantaneous twisting force on the flywheel, commensurate with the torque imposed by sudden loads. Load-induced torque slows and, in extreme cases, stalls out the engine. The relationship between engine torque and load torque is known as the torque rise.

$$\text{Torque rise \%} = \frac{(\text{peak torque demand} - \text{rated engine torque} \times 100)}{\text{rated engine torque}}$$

If peak torque demand were twice that of engine torque, the torque-rise percentage would be only 50% and we could expect the engine to stumble and stall under the load. If, on the other hand, engine torque equaled load-induced torque, the engine would absorb the load without protest. However, high-torque-rise engines stress drivelines, mounts, and related hardware. Some compromise must be made.

It should also be noted that not all driven equipment imposes sudden torque rises. For example, centrifugal pumps and blowers cannot lug an engine because the efficiency of these devices falls off more quickly with reduced speed than engine torque. Gen-sets run at constant speed and do not require much by way of torque rise. On the other hand, positive-displacement pumps generate high torques when pump output is throttled.

As the engine slows under load, the governor increases fuel delivery. Naturally aspirated engines respond quickly, since the air necessary to burn the additional fuel is almost immediately available. Turbocharged engines exhibit a perceptible lag as the turbo spools up. But naturally aspirated engines have difficulty in meeting emissions limits and, for the same power output, are heavier and more expensive than turbo-charged models. Some industrial engines employ a small turbocharger for low-speed responsiveness and a second, larger unit for maximum power. Variable geometry tur-bocharging (VGT) virtually eliminates lag time.

The black smoke accompanying turbo lag can be reduced with an air/fuel ratio controller. Also known as a smoke limiter, the device limits fuel delivery until suf-ficient boost is present for complete combustion. Adjustment is a trade-off between transient smoke and engine responsiveness.

The engine manufacturer normally has final say on the mounting configuration that may consist of parallel rails or, in the case of several Caterpillar oilfield engines, a com-pound base. In the later arrangement the engine bolts on an inner base, which is suspended on springs above the outer base. The technician has the responsibility to see that engine mounts permit thermal growth and that engine and driven element are in dead alignment.

Thermal expansion

Cast iron "grows" less than steel when exposed to heat. The coefficient of expan-sion is 0.0000055 for cast iron and 0.0000063 for steel. A 94-in.-long iron engine block will elongate 0.083 in. as its temperature increases from 50° to 200°F. Under the same temperature increase, 94-in. steel mounting rails grow 0.089 in. These parts must be free to expand.

Caterpillar 3508, 3512, and 3516 oilfield engines mount on a pair of factory-supplied rails bolted to the oil pan. Standard procedure is to tie the engine down to the rail with a fitted bolt, that is, a bolt inserted with a light push fit into a reamed hole at the right rear corner of the oil pan. This bolt provides a reference point for alignment. Other engine-to-rail mounting bolts fit into oversized holes in the rails to allow for expansion. Clearance-type bolts should be 0.06 in. smaller than the diam-eter of the holes in the rails (Figure 3-2). Chocks, used as an installation convenience to position the engine on the rails, must not constrain thermal expansion.

Alignment

Begin by cleaning all mating surfaces to remove rust, oxidation, and paint. If rub-ber couplings are present, remove them. It may be necessary to fabricate a dial-indicator holder from 1½-in. steel plate that can be bolted down to the machine. While this

ALIGNMENT CHOCKS
(REAR AND FRONT
SIDE ONLY)

MOUNTING BLOCK (4)
AND SHIMS

CLEARANCE
TYPE BOLTS

DO NOT USE FRONT
RAIL RESTRAINT

3-2 General arrangement indicating chock and shim positions for mounting a stationary engine on rails. Courtesy Caterpillar Inc.

sounds like overkill, many commercially available magnetic indicator holders lack rigidity. Flex in the holder can be detected by the failure of the indicator to return to zero when the shaft is rotated back to the initial measurement position.

Parallel misalignment occurs when the centerlines of the engine crankshaft and driven equipment are parallel, but not in the same plane (Figure 3-3). Mount a dial indicator on the engine flange with the point against the driven flange. Make several readings while a helper bars over the crankshaft in the normal direction of rotation.

The driven, or load, shaft should, as a general rule, be higher than the engine shaft. Engine main bearings typically have greater clearance than driven-equipment bearings. Until the engine starts, the crankshaft rests on the main-bearing caps, one or two thousandth of an inch below its running height. In addition, some allowance must be made for vertical expansion of the engine at operating temperature, which is nearly always more pronounced than the vertical expansion of the driven machinery.

Figure 3-4 illustrates a method of verifying dial-indicator readings on the outer diameters (ODs) of flanges and other circular objects. Zero the indicator at A in the drawing and, as the flange is rotated, make subsequent readings 90° apart. If the indicating surface is clean and the instrument mount secure, the needle will return to zero in the A position and B + D readings will equal C.

Measurement of angular misalignment can be made with a feeler gauge (Figure 3-5). Once the problem is corrected by shimming, bolt the flanges together and verify that the crankshaft can move fore and aft a few thousandths of an inch against its thrust bearing.

3-3 Parallel misalignment occurs when the centerlines of drive and driven equipment are parallel, but not in the same plane. Courtesy Caterpillar Inc.

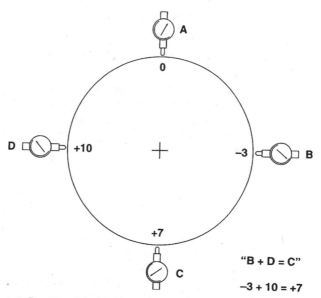

3-4 B + D = C holds for round objects measured with accurate dial indicators. Courtesy Caterpillar Inc.

3-5 Angular misalignment nearly always reflects the lay of the shafts. But mis-machined flanges can also contribute to the problem. Courtesy Caterpillar Inc.

Parallel and angular misalignment can originate in driveline hardware, which rarely gets the scrutiny it deserves (Figure 3-6). Parallel misalignment, called bore runout, refers to the lack of concentricity between the bore of a hub and the shaft centerline. Angular misalignment occurs when the mating face of a flange is not perpendicular with the shaft centerline. This sort of machining error is known as face runout.

Bore runout between the flywheel inner diameter (ID)) and the crankshaft pilot-bearing should be no more than 0.002 in., and no more than 0.005 in. for adaptors that make up to the flywheel. The maximum face runout for driveline components is 0.002 in.

Generators, transmissions, and other close-coupled driven elements that bolt directly to the flywheel housing can also suffer from misalignment. As a check, loosen the bolts securing the unit to the flywheel housing and measure the gap between the parts with a feeler gauge. The interfaces should be within 0.005 in. of parallel.

It is good practice to determine the bending moment of components cantilevered off the flywheel housing, as described previously under truck "Motor mounts."

Use brass shims (steel shims expand as they rust) and tighten the typical four-bolt mounting as shown in Figure 3-7. Grade-8 hold-down bolts are torqued to specification, an operation that requires a large torque wrench and a torque multiplier in the form of a 3- or 4-ft cheater bar.

Marine engines

Because of the one-off nature of most small-boat construction, the technician is very much on his own when mounting an engine. Figure 3-8 provides an overview of the installation of Yanmar engines on sail and powered boats.

CLUTCH COMPOUND

SIDE VIEW

**BORE MISALIGNMENT
UP/DOWN**

TOP VIEW

**BORE MISALIGNMENT
RIGHT/LEFT**

SIDE VIEW

**FACE MISALIGNMENT
UP/DOWN**

TOP VIEW

**FACE MISALIGNMENT
RIGHT/LEFT**

3-6 Possible face and bore misalignments. Courtesy Caterpillar Inc.

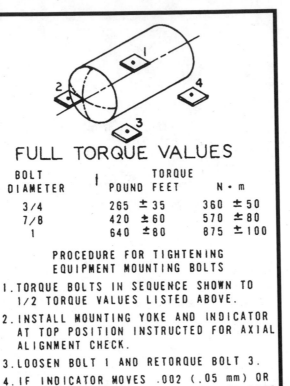

FULL TORQUE VALUES

BOLT DIAMETER	TORQUE	
	POUND FEET	N·m
3/4	265 ±35	360 ±50
7/8	420 ±60	570 ±80
1	640 ±80	875 ±100

PROCEDURE FOR TIGHTENING EQUIPMENT MOUNTING BOLTS

1. TORQUE BOLTS IN SEQUENCE SHOWN TO 1/2 TORQUE VALUES LISTED ABOVE.

2. INSTALL MOUNTING YOKE AND INDICATOR AT TOP POSITION INSTRUCTED FOR AXIAL ALIGNMENT CHECK.

3. LOOSEN BOLT 1 AND RETORQUE BOLT 3.

4. IF INDICATOR MOVES .002 (.05 mm) OR LESS RETORQUE BOLT 1 AND FOLLOW STEP 5 AND 6. IF INDICATOR MOVES MORE THAN .002 (.05 mm) ADD SHIMS UNDER BOLTS 1 OR 3. LOOSEN ALL BOLTS AND REPEAT STEPS 1 THRU 4.

5. LOOSEN BOLT 2 AND RETORQUE BOLT 4.

6. IF INDICATOR MOVES .002 (.05 mm) OR LESS, RETORQUE BOLT 2. IF INDICATOR MOVES MORE THAN .002 (.05 mm) ADD SHIMS UNDER BOLTS 2 OR 4. REPEAT STEPS 1 THRU 6.

7. WITH INDICATOR AT TOP RETORQUE ALL BOLTS TO FULL VALUES. READING SHOULD NOT CHANGE MORE THAN .002 (.05 mm).

3N2021

3-7 Hold-down bolt torque limits and tightening procedure. Courtesy Caterpillar Inc.

3-8 Engine-room layout and driveline configurations for Yanmar and other small engines.

Table 3-1. Propeller-shaft angles relative to the waterline

Drive type	Recommended angle	Max. permissible angle
Parallel drive	10°	15°
Angle drive	0°–3°	8°
V-drive	5°–8°	15°

Mounts

Inboard engines should be mounted level port-to-starboard and inclined no more than 15° off the horizontal (Table 3-1). Higher crankshaft angles reduce drive efficiency and increase the likelihood of oil starvation for engines with forward mounted oil-pump pickup tubes.

Your supplier will assist in determining propeller diameter, pitch, and style. As installed, the propeller should be at least one propeller diameter below the waterline and permit the engine to achieve maximum rated speed under full load at wide-open throttle. If the engine overspeeds by more than 5%, the propeller is undersized.

Once you have determined the inclination angle at which the engine will be installed, construct the bed, which should have as large a footprint as hull construction permits. Leave room around the engine for servicing and, if possible, arrange matters so that the engine can be removed without structural alterations to the vessel. There must also be sufficient clearance between the flywheel cover/transmission and hull for compression of the flexible motor mounts.

Determine the height of the crankshaft centerline with the engine resting on its mounts. You should be able to get this data from the engine maker. The next and very critical operation is to mark holes for the prop shaft. Some yards use a laser alignment tool, while others rely upon the more traditional approach described here.

Construct a wooden jig, like the one shown in Figure 3-9, and secure it to the forward engine-room bulkhead. Run a string from the jig to the stern box. Verify that the inclination is correct and that the string is on the vessel centerline. Measure the vertical distance between the string and hull to establish that the transmission, clutch, and other components have sufficient clearance.

Using the string as a guide, mark stern-tube hole locations on the forward and aft sides of stern, or stuffing, box. Drill small holes where indicated on both sides of the box. Pass the string through these holes, which almost certainly will not center in the previously drilled holes. Figure 3-10 illustrates how the final hole dimensions are arrived at. Hole diameter should exactly match stern-tube diameter.

At this point, the crankshaft centerline and stern tube should be in fairly close alignment. Seal the stuffing-box holes with silicone, bolt down the stern tube, and secure the free end of the string to the center of the prop shaft. Using a fabricated marker, align the shaft with the string (Figure 3-11).

The next step is to fabricate an engine-mounting jig with holes for the string at crankshaft location (Figure 3-12). Place the jig on the engine bed and adjust the height and position of the mounts as necessary.

Fix to partition wall of engine room.

Plate A

Centering line

Detail

Plate A

Back side

3-9 An alignment jig mounts on the engine-room forward bulkhead.

Finally, mount the engine and align the prop-shaft flange with the transmission flange as described previously for stationary engines.

Exhaust systems

Wet exhaust systems reduce engine-room heat and simplify installation by allowing oil- and heat-resistant flexible hose to be used aft of the mixing elbow (Figure 3-13). Basic specifications are:

- Exhaust outlet at transom—at least 6 in. above the loaded draft line.
- Elbow inboard of the transom—at least 14 in. higher than the draft line.

3-10 Accurate alignment is a process of refining approximations: the boat builder drills pilot hole through both faces of the stern box and threads the string through it for a better fix on shaft alignment.

3-11 In preparation for final alignment, a home-made indicator is affixed to one of the flange bolt holes and the shaft centered over the string.

3-12 A jig that replicates the crankshaft-centerline/motor-mount relationship helps to avoid the embarrassment of redrilling engine-mount holes. Once the shaft aligns with the jig, all that remains is to verify flange alignment as described for stationary engines.

3-13 Exhaust system showing major elements and dimensions above the loaded draft line. The draft line is a convenient marker, but the critical reference is the waterline. Power boats assume a stern-down attitude at speed, which raises the water line. In the application shown, a riser is used to elevate the mixing elbow and the water-injection point 6 in. or so above the waterline.

- Muffler—inboard of the elbow, with the outlet inclined about 8° off the horizontal.
- Water trap—located at the lowest point of the system between the muffler and engine. While a water trap is recommended for all installations, the backpressure developed by the unit can result in turbocharger surge. If this condition occurs, consult the engine manufacturer. It may be necessary to derate the engine by fitting a smaller propeller.
- Cooling-water injection point—6 in. or more above the draft line to prevent water in the hose from flowing back into the engine upon starting and after shutdown. When necessary, a riser can be used to raise the height of the mixing elbow.

4
CHAPTER

Basic troubleshooting

This chapter provides basic troubleshooting information that should narrow the source of the problem to the fuel, air, engine mechanical, starting or exhaust systems. Once this is done, turn to the chapter that deals with the system for additional troubleshooting and repair information.

Until about 30 years ago, diesel engines were purely mechanical objects and relatively easy to diagnose. Modern diesels are electro-mechanical hybrids, almost robotic in the way they function. Engine or electronic management systems (EMS) regulate fuel timing and volume, turbo boost, engine rpm and, in short, every critical engine function. The onboard computer does this by integrating sensor data about engine operation with environmental data and operator throttle commands. While these management systems have civilized the diesel engine, smoothing its rough spots, cleaning its exhaust and dramatically improving its power output, the price has been increased complexity and greater opportunity for failure. A bad sensor, a sticking actuator or a bent pin in an electrical connection can shut the engine down. Most of these failures are not visible to the eye.

Diagnostic trouble codes (DTCs) compensate in part for the complexity. The first step when diagnosing computerized engines is to retrieve the trouble codes with a scan tool. The second step is to gather enough knowledge about the system to understand how it works and what the retrieved DTCs mean. The onboard computer speaks in generalities that require interpretation. Chapter 6 describes the functioning of these systems in some detail with Caterpillar HEUIs and Ford Power Strokes receiving special attention. But engine management systems have undergone constant development as computers became more powerful and emissions limits more rigorous. There is no universal, one-size-fits-all EMC. And, of course, computerized engines are also subject to mechanical failures. Injectors carbon over, high-pressure pumps fail, and gaskets blow. Many of these mechanical problems do not trigger a trouble code.

Again, let me stress that the first step in diagnosing EMS engines is to read the trouble codes. If you merely want to keep an eye on the condition of your engine, a $100 scanner that retrieves the codes and erases them with a push of the button is adequate. Some malfunctions require professional equipment and access to factory documentation to diagnose.

A $3000 scanner will read active DTCs—those in effect now—and inactive, or historical, trouble codes that have been set and later erased. The tool will also retrieve events a few seconds before and after each trouble code was set, shut down individual injectors, and report the real-time data sensors transmit to the computer. Amateur mechanics do what they can with the tools they have and farm out the really hard problems to professionals.

Malfunctions

Common malfunctions are listed below together with probable causes and remedial actions. Fuel-related malfunctions can affect the color of the exhaust smoke, as indicated in Table 4-1. Catalytic converters and diesel particulate matter filters (see Chap. 13) must be disconnected for the engine to produce visible exhaust smoke. Taken together, engine behavior and smoke color provide a good indication of the source of the problem.

Engine cranks slowly, does not start

Starting system malfunction	Recharge batteries if cranking voltage drops below 9.5V or electrolyte reads less than 1.140 when tested with a hydrometer. Clean battery terminals. Perform starting system check (Chap. 11).
High parasitic loads	Check for binds in driven equipment, overly tight drive belts, shaft misalignments.
Crankshaft viscosity bound	Dark, sticky residue on the dipstick can indicate presence of antifreeze (ethylene glycol) in oil. Have oil analyzed.

Engine cranks normally, does not start

EMS sensor or actuator failure	Retrieve trouble codes. See Chap. 6.
No fuel to injectors	Check for restrictions or air leaks in fuel system, low injector pressure.
Contaminated fuel	Flush fuel system.
Glow-plug failure	Check glow-plug supply circuit, glow-plug control module and individual plugs.
Air inlet restriction	Replace air filter element.
Exhaust restriction	Inspect piping, if necessary perform backpressure test.
Loss of cylinder compression	Test cylinder compression.

Engine starts normally, runs no more than a minute or two, and shuts down

Intermittent EMS-related failure, bad harness connection	Retrieve trouble codes. See Chap. 6.
Air in fuel	Bleed system and check for air leaks.
Fuel return line restricted	Disconnect line to verify flow and remove obstruction.
Air inlet restriction	Replace filter element.
Clogged fuel filter	Replace filter element.

Engine starts normally, misfires

EMS-related failure	Retrieve trouble codes. See Chap. 6.
Air in fuel	Bleed system, check for air leaks.
Air inlet or exhaust system restriction	Replace air filter, check for excessive backpressure.
Clogged fuel filter	Replace filter element.
Malfunctioning injector(s)	Replace injector(s).
Loss of compression in one or more cylinders	Check cylinder compression.

Engine fails to develop normal power

EMS-related fault	Retrieve trouble codes. See Chap. 6.
Insufficient fuel supply	Replace filters, check transfer pump output, cap vent, and air in the system.
Contaminated fuel	Verify fuel quality.
Injection timing error	Check injector-pump timing.
High-pressure fuel system malfunction	Inspect high-pressure fuel system for leaks and air entrapment. Verify pump output pressure. If necessary have pump recalibrated.
Air inlet restriction	Change air-filter element.
Exhaust restriction	Inspect exhaust piping, test for excessive backpressure.
Insufficient turbo boost	Change air-filter element, check boost pressure and for exhaust leaks upstream of the turbo.
Loss of engine compression	Check engine cylinder compression.

Table 4-1. Diagnosis by exhaust smoke color

Black or dark gray smoke Symptom	Probable cause	Remedial action
Smokes under load, especially at high and medium speed. Engine quieter than normal.	Injector pump timing retarded.	Set timing.
Smokes under load, especially at low and medium speed. Engine noisier than normal.	Injector pump timing advanced.	Set timing.
Smokes under load at all speeds, but most apparent at low and medium speeds. Engine may be difficult to start.	Weak cylinder compression.	Repair engine.
Smokes under load, especially at high speed.	Restricted air cleaner.	Clean/replace air filter element.

(*Continued*)

Table 4-1. Diagnosis by exhaust smoke color (*Continued*)

Blue, blue-gray, or gray-white smoke Symptom	Probable cause	Remedial action
Smokes under load, noticeable loss of power.	Turbocharger malfunction.	Check boost pressure.
Smokes under load, especially at high and medium speeds. Power may be down.	Dirty injector, nozzle(s).	Clean/replace injectors.
Smokes under load, especially at low and medium speeds. Power may be down.	Clogged/restricted fuel lines.	Clean/replace fuel lines.
Puffs of black smoke, sometimes with blue or white component. Engine may knock.	Sticking injectors.	Repair/replace injectors.
Whitish or blue smoke at high speed and light load, especially when engine is cold. As temperature rises, smoke color changes to black. Power loss across the rpm band, especially at full throttle.	Injector pump timing retarded.	Set timing.
Whitish or blue smoke under light load after engine reaches operating temperature. Knocking may be present.	Leaking injector(s).	Repair/replace injector(s).
Blue smoke under acceleration after prolonged period at idle. Smoke may disappear under steady throttle.	Leaking valve seals.	Replace seals, check valve guides/stems.
Persistent blue smoke at all speeds, loads and operating temperatures.	Worn rings/cylinders.	Overhaul/rebuild engine.
Light blue or whitish smoke at high speed under light load. Pungent odor.	Over-cooling.	Replace thermostat.

Tests

Fuel quality

Take a fuel sample from some convenient point upstream of the filter/water separator. Allow the fuel to settle for a few minutes in a glass container and inspect for cloudiness (an indication of water), algae (jellylike particles floating on the surface), and solids. Placing a few drops of fuel between two pieces of glass will make it easier to see impurities. If fuel appears contaminated, drain and purge the system as described below.

Assuming that the fuel is clean and water-free, the most important quality is its cetane value, which can be determined with fuel hydrometer. A good No. 2 diesel has a specific gravity of 0.840 or higher at 60°F. No. 1 diesel can reduce power outputs by as much as 7% over No. 2 fuel. Blending the two fuels, a common practice in cold climates, results in correspondingly less power.

Fuel system

Most failures involve the fuel system. This system consists of three circuits, each operating at a different pressure as mentioned below:

- *Low-pressure circuit.* This includes the tank strainer, in-tank pump (on many vehicles), water-fuel separator, filter(s), and lift pump. Pressures vary with the application, but rarely exceed 75 psi.
- *Fuel-return circuit.* Operating at almost zero pressure, the return line conveys surplus fuel from the injectors back to the tank, filter, or to the inlet side of injector pump. Many of these systems include a restrictor orifice between the fuel-supply line and the return line. The orifice can clog, closing off the fuel return. Air leaks become a matter of concern when fuel is returned to the filter or injector pump, and not to the tank.
- *High-pressure circuit.* As shown in Figure 4-1, the high-pressure circuit connects the discharge side of the injector pump with the injectors. For many applications, the connecting plumbing takes the form of dedicated lines from pump to the individual injectors. A more modern approach is to supply the injectors from a common manifold, or rail. The third alternative does away with the external high-pressure circuit by employing unit injectors (UIs). Each UI is a self-contained unit, with its own high-pressure pump operated by an engine-driven camshaft or, in the case of the Hydraulic Electronic Unit Injector (HEUI), by oil pressure. The unit pump system (UPS) is a hybrid, with each injector served by a dedicated pump plunger operating off the engine camshaft. Plunger-to-injector lines are pressurized.

A mechanic has a duty to himself to know the pressures he is dealing with. Most, but not all, older-model engines had fuel systems that operated in the neighborhood of 6000–12,000 psi. But even at 6000 psi, fuel easily penetrates the skin (witness the speckled hands of old diesel mechanics) and often results in blood poisoning. Common-rail and other modern systems generate pressures on the order of 30,000 psi. Pressures of this magnitude cut to the bone and, if air bubbles are present, the fuel jets out like water from a hose.

WARNING: Wear eye protection and heavy gloves when working on diesel fuel systems of any vintage. Do not attempt to disable modern ultrahigh-pressure injectors by cracking fuel-line connections. High-pressure fuel leaks may not be visible, but can be detected with as splatter on a piece of cardboard placed next to the connection.

Diesel-Einspritzsysteme von Bosch

VP44
Radialkolben-
Verteiler-
einspritzpumpe

CRS
Common-Rail-
System

UIS
Unit-Injector-
System

UPS
Unit-Pump-
System

4-1 This drawing illustrates the high-pressure circuits for four modern fuel systems. From a mechanic's point of view, the UI system has much to recommend it. High pressures are confined within the injector bodies and failures tend to be cylinder-specific. (Photo Bosch)

Hydraulically actuated HEUI injectors shift the high-pressure regime to lube oil supplied at between 800 and 3300 psi, which is still a considerable pressure. These and other electronic injectors (recognized by the presence of wires running to them) pose the risk of electroshock. Injector voltages and amperages can be lethal.

WARNING: Do not disconnect the wiring to electronic injectors while the engine is running.

For fuel to enter the cylinders, the h-p pump must generate enough pressure to unseat the injectors. Pop-off pressure, known more formally as NOP (nozzle opening pressure) or VOP (valve opening pressure), depends upon an unrestricted fuel supply. Air leaks or fuel blockages upstream of the pump, or failure of the pump itself can reduce delivery pressure below NOP. Electronic engine management systems keep close watch on delivery pressure and its effect on other variables, such as the concentration of oxygen in the exhaust. Low delivery pressure triggers one or more trouble codes.

Older, precomputer engines require a more proactive approach to determine if fuel is getting to the cylinders. One technique is to spray starting fluid into the air intake while cranking. If the engine starts, runs for few seconds and stalls, you can be reasonably confident that the problem lies in the fuel system.

CAUTION: Employ starting fluid with discretion. Glow plugs can ignite the fluid in the manifold and large amounts of starting fluid in the intake manifold or in the scavenging system on two-cycle engines can result in explosions powerful enough to break piston rings and bend connecting rods.

An alternative approach is to crank the engine over for a 20 or 30 seconds, and crack the fuel-supply line to one or more injectors. Fuel should be present. Note that this technique should be confined to older engines known to generate moderate fuel pressures of 6000 psi or so.

Bleeding

Bleeder screws or Schrader valves are located at high points in the filter assemblies (Figure 4-2) and in one or more locations on the accessible side of most injector pumps (Figure 4-3). Carefully clean the screws and adjacent areas to prevent foreign matter from entering the fuel circuit. Many engines have a hand-operated bleeder pump as part of the lift pump; others must be primed by cranking. A portable hand-pump can be used for engines not equipped with bleeder pumps.

Build a few psi of pressure in the system and, working from the tank forward, loosen the bleeder screws. Tighten the screws as soon as fuel flows in an uninterrupted stream.

4-2 Fuel filter bleed screws. Lehman Ford Diesel.

1 Injector pipe clamp
2 Union nut
3 Injector pipe clamp
4 Delivery valve holder
5 Valve holder clamp
6 Bleed screws

4-3 Inline injection pump bleed screws. Ford Industrial Engine and Turbine Operations.

Run the engine for at least 20 minutes to ensure that the fuel system is completely purged. If the no-start or hard-start problem recurs after the engine has cooled down, the problem is almost surely an air leak in the low-pressure (injector-pump-supply) circuit.

Low-pressure circuit tests

Failure is most often associated with flow restrictions and air or fuel leaks, although failure of the lift pump is not unusual.

The tests that follow, developed by Ford for light trucks, locate flow restrictions by measuring the pressure drops across individual components while the engine is running. In other words, Ford uses the engine as a test bench. Other manufactures would have the mechanic disassemble the system and test each component separately. If you want to use the Ford approach—which has advantages—establish baseline pressure drops while the system is still healthy and operating normally.

Which gauge will be used depends upon the fuel-filter location relative to the lift pump. If the filter is located on the suction side of the lift pump, a 0–30 in. Hg vacuum gauge will be needed. If downstream of the pump, the Ford application calls for 0–15 psi pressure gauge. Other lift pumps develop higher pressures.

1. Connect the appropriate gauge at the output side of the filter (Figure 4-4). Start the engine and note the gauge reading at 2500 rpm.
2. Connect the gauge to the inlet side of the filter, repeat the test, and compare these readings against the pressure drop for a new filter (Figure 4-5).
3. Use the same procedure for detecting restrictions in the water/fuel separator, heater, and other components.

4-4 Filter location relative to the lift pump determines whether a pressure or vacuum gauge is used. In this example, the filter is on the outlet side of the pump and sees positive pressure.

4-5 The next step is to install the gauge at the input side of the filter. The difference between output-side and input-side readings represents the pressure drop across the filter.

The lift pump, also known as the transfer or supply pump, moves fuel from the tank to the injector pump. Figure 4-6 illustrates the Ford approach to measuring lift-pump output pressure. A container of diesel fuel supplies the pump, thus isolating it from possible suction-side restrictions. However, it is usually enough to connect the gauge in parallel with pump output.

4-6 When measuring lift-pump output, isolate the pump from suction-side pressure drops by providing fuel from a separate source.

Pump pressure is only part of the picture. Manufacturers should, but rarely do, provide volume specifications for this and other pumps. If you're familiar with the engine, some rough idea of output volume can be had by monitoring fuel flow leaving the return line.

Fuel leaks leave a trail, but some mechanics like to check connections with soapy water or the glycerin-based solutions used to detect Freon leaks.

Air can enter on the suction side of the lift pump or when voids develop in the solid column of fuel. The latter condition occurs if the tank runs dry, filter elements are changed or, after shutdown, when fuel cools and contracts in filter canisters.

Entrapped air causes the injector pump to lose prime. Symptoms include reluctance or outright refusal to start, and sudden shutdowns within seconds of starting. Depending upon the amount of air intrusion, the engine may start more or less normally, but refuse to idle.

The time required for air to reach the injector pump gives an indication of the leak source. If the engine dies within seconds of starting, the source of the air must be nearby. Layouts vary, but common practice is to mount the filter and water separator in close proximity to the injector pump. Check these items first. Fuel-return lines that discharge into the filter or suction side of the injector pump present a major leak hazard.

Another technique is to insert a length of clear plastic fuel line of the type used on motorcycles between the lift and injector pumps (Figure 4-7) or at any location in the low-pressure circuit where leaks are suspected. Start the engine and look for bubbles in the fuel stream.

4-7 A length of clear plastic tubing spliced into the fuel line between the lift and injection pumps shows if air is present.

You can also test for air leaks by disconnecting the line at the tank and pump. Plug one end and connect a hand pump or other source of low-pressure air to the free end. Pressurize to no more than 15 psi and apply soapy water or a commercial leak detector solution to the line connections, filter-canister parting surfaces, and other potential leak sources.

CAUTION: Do not pressurize the fuel tank—it might rupture.

Once the leak has been found and repaired, renew o-ring seals on connections, assemble all threaded connections with Loctite 515 Gasket Eliminator, and bleed the system as described above.

High-pressure circuit tests

High-pressure systems fail because of fuel leaks (air leaks are uncommon), malfunctioning injectors, or inadequate pump pressure.

Fuel-pipe unions on engines with pump-fed injectors are a major leak source. Loosening and retightening the union usually corrects the problem. See Chap. 5 for additional information.

Some engines are more prone to leakage than others. For example, hard-used White trucks tend to develop small leaks at the fuel delivery valves, located under the fuel-pipe connections on UTDS, APD, 6BB, T and Q series injector pumps. Replace the o-ring seal and copper ring gasket, and inspect lapped surfaces for channeling. The UTDS pumps in question develop more than 15,000 psi. Crimps or flattened spots in the fuel lines, over-fueling, and clogged injectors will further increase delivery pressure.

WARNING: Wear eye protection and keep hands clear of high-pressure fuel spray. Use a piece of cardboard to detect leaks in hard-to-see places.

Current EM systems focus on overall performance, rather on events occurring within individual cylinders. These systems know when a cylinder goes down or when electronic injectors lose power, but are blind to the roughness and loss of responsiveness associated with skewed spray patterns, fuel dribble, and the like. For that we need more traditional methods.

Symptoms of injector failure include:

- Ragged idle or misfiring at certain speeds
- Black, gray, or white smoke
- Vibration and knock

If the injector fails completely, the exhaust runner for that cylinder will be noticeably cooler than the others immediately after startup. Glow-plug resistance, which increases with temperature, will be lower for the affected cylinder. On some engines with remote injector pumps, it is possible to feel the sudden contraction of the fuel pipe as a healthy injector snaps open. The piping to a malfunctioning injector feels inert by comparison. A more direct approach is to disable suspect injectors. This is done by depressing unit-injector cam followers with a pry bar, cracking the fuel-inlet connections on low-pressure pump-fed injectors, or by denying power to solenoid or piezo injectors. A good scan tool will disable electronic injectors safely and on demand. If the injector has failed, taking it out of action has little or no effect upon engine rpm.

The best way of verifying injector performance is to substitute known good injectors and observe the effect on idle, smoothness, and smoke production.

As described in Chap. 5, mechanical injectors can often be repaired in the field. Repairs of electronic injectors are best left to specialists who have the necessary test equipment and parts inventory.

The classic symptom of injector pump failure is low delivery pressure. If the engine is to start, NOP must be attained at cranking speed. If it is to develop maximum torque, the pump must be capable of delivering its full rated pressure and volume. Low or no pump pressure on computer-controlled engines is most often the result of a failed crankshaft position sensor. The throttle position sensor is the next most likely suspect.

Precomputer pumps are, for the most part, rpm-dependent. Test output pressure by connecting a suitable gauge to the test port, usually found at the discharge end of the pump (Figure 4-8) or at the pump side of the return-line restrictor. Many of these engines also include a key-operated fuel shutoff downstream of the pump. A typical example is shown in Figure 4-9. And most modern pumps include a bellows or diaphragm-operated altitude-compensator mounted on the pump body (see Chap. 5 for more information). A pin-hole leak in the diaphragm can mimic pump failure.

The governor section of mechanical pumps includes an array of adjustment screws, most of which are sleeping dogs, better left undisturbed. The two that a mechanic needs to know about, set the idle speed and maximum governed speed, usually by restricting the movement of the throttle control lever. The adjustments shown in Figure 4-10 are fairly standard, although more elaborate designs sometimes split idle speed into two ranges.

4-8 Internal-pressure test port for a Stanadyne distributor-type pump.

During operation   At stopping

4-9 solenoid-operated fuel shutoff valve is a standard feature on modern pumps, some of which also incorporate an air shutdown.

4-10 Idle and high-speed adjustment screws as found on several Yanmar pumps.

Air inlet system

In its most highly developed form, the air inlet system incorporates a turbocharger and exhaust gas recirculation (EGR).

Air filter

Symptoms of filter restrictions are:

- Hard starting
- Black exhaust smoke
- Loss of power
- Failure to reach governed speed
- Erratic idle (naturally aspirated engines)
- Turbocharger speed surges and possible oil pullover

Service manuals for vehicle and other small engines merely provide a schedule for changing the air filter. If the operating environment varies, as it does on construction sites or during off-road operation, a better approach is to change the element when the pressure drop across it becomes excessive. Some experimentation with new and used filter elements will be necessary to establish a standard. Pressure drops across new filter elements vary over a range of 2–15 in./H_2O. As a general rule, a 50% increase in pressure drop is cause enough to replace the element.

Figure 4-11 illustrates the hookup for measuring pressure drop. The test is made with the engine running at the speed that generates maximum air flow. Air flow peaks at fast idle for naturally aspirated engines and at full power for turbocharged engines.

CAUTION: Do not operate the engine with the air cleaner removed. Unlike spark-ignition (SI) manifolds, which are obstructed by venturis and throttle plates, diesel manifolds open directly (or via expensive compressor wheels) to the combustion chambers. A dropped wing nut or, in the case of turbocharged engines, careless fingers, will have dramatic effects.

Positive crankcase ventilation (PCV)

Positive crankcase ventilation valves require routine replacement, since cleaning is rarely effective. The 2.2-L Isuzu used in several GM vehicles demonstrates how complex these systems can become. The liquid component of blowby gases is filtered out and collected in a holding tank for return to the crankcase. A check valve under the tank isolates the sump from PCV vacuum. If this valve clogs, the oil level rises in the tank and is pulled over into the intake manifold to produce clouds of blue smoke.

Exhaust gas recirculation

Failure of the exhaust gas recirculation (EGR) valve or control circuitry flags trouble codes, increases smoke levels dramatically and costs power. When used with EMS, the EGR valve incorporates a position sensor and an actuator that can be cycled with the appropriate scanner.

Turbocharger

What follows is a quick rundown on some of the more frequently encountered turbo faults. See Chap. 9 for more information about these complex machines.

Atmospheric pressure for comparison

Inlet

Air cleaner

Air box

Air flow

Filter causes pressure drop

4-11 Stationary and marine technicians sometimes used a manometer—a U-shaped tube, partially filled with water, with one leg open to the atmosphere and calibrated in quarter-inch increments—to ·measure the pressure drop across the filter. Field mechanics prefer the convenience of a vacuum gauge. The Ford gauge is calibrated in inch/H_2O.

Turbocharger failure has a number of causes, most of them associated with the demands put on the lubrication system by these devices that, in their most modern form attain as much as 400,000 rpm under peak load (Figure 4-12). Momentary interruptions in the oil supply, which might occur during initial startup after an oil change or during sudden acceleration immediately after a cold start, destroys the bearings.

Turbine wheel

Turbine exhaust gas outlet

Exhaust gas discharge

Engine cylinder

Compressor

Exhaust gas inlet

Compressed air discharge

Inter cooler

Ambient air inlet

Ambient air inlet

Compressor wheel

➡ Engine exhaust gas flow

⇨ Compressed air flow

Turbine

4-12 Gas flow through a Detroit Diesel turbocharger.

If the engine is run hard and abruptly shut down, oil trapped in the turbocharger case absorbs heat from the turbine and carburizes into an abrasive. Many EM systems can be programmed to eliminate cold, wide-open-throttle starts and hot shutdowns.

Other problems include carbon and scale accumulations on the wheels, exhaust-side leaks (caused by thermal expansion of the piping), pressure-side leaks (often at the manifold-to-block gasket), and blade-tip erosion. A faulty air filter will "dust" the compressor wheel, giving it a satiny appearance.

Some malfunctions—noise, excessive bearing clearances, cracked exhaust plumbing—obviously involve the turbocharger. A fall off in power, sooty exhaust, low or erratic turbo boost may have causes outside of the unit. Before assuming that the turbocharger is a fault, check the fuel system with particular attention to the injectors, and verify that the pressure drop across the air filter element is within specification. Carefully inspect the turbocharger oil and vacuum lines. And, of course, you need to retrieve the trouble codes. Many EMS blowers incorporate variable geometry, so that boost comes on early during the rpm curve. Malfunctions will flag trouble codes associated with abnormal levels of boost and actuator failure.

Start the engine and listen to the sound the turbocharger makes. With a little practice you will be able to distinguish the shrill sound of air escaping between the compressor and engine and the sound of exhaust leaks, which are pitched at a lower register. If the sound changes in intensity, check for a clogged filer, loose sound-deadening material in the intake duct, and dirt/carbon accumulations on the compressor wheel and housing. A stethoscope will amplify any mechanical noise.

Glow plugs

Slow cold starts and persistent white smoke can be caused by failure of one or more glow plugs on indirect injection (IDI) engines. Test individual plugs with the ignition off and a low-voltage ohmmeter connected between the plug terminal and a good engine ground. Most exhibit about 2 Ω hot, and nearly 0 Ω cold. If the heating element is open, resistance will be infinite.

An open supply circuit denies power to all glow plugs and makes cold starting virtually impossible. Check for blown fusible links verify that the relay, often a solid-state device with an internal resistance that drops glow-plug voltage to 6V, is functional. EMS glow-plug controllers vary plug on-time with ambient and coolant temperatures. See Chap. 11 for details.

Exhaust backpressure

Excessive backpressure can prevent starting, cost power, increase exhaust temperature, and color the exhaust. Water locks—exhaust pipe risers to block water entry on marine installations—should not be used with turbocharged engines.

Look for damaged mufflers and flattened, crimped, or improperly sized exhaust piping. Trouble codes should be flagged if the EMS senses a clogged particulate trap or catalytic converter.

If excessive backpressure is suspected and the source is not obvious, drill and tap a 11/32 in. hole on a straight section of exhaust piping 6–12 in. downstream of the turbocharger. Tap threads for a 1/8 in. pipe fitting. Connect a manometer and, with coolant temperature normal, apply the maximum possible load to the engine. While backpressure varies with the installation, readings higher than 3 in./Hg are cause for concern.

Engine mechanical

Engine malfunctions fall into four categories: fluid leaks, excessive oil consumption, loss of compression, and bearing wear.

Fuel leaks

External fuel leaks are discussed earlier in this chapter. Fuel contamination of the lube oil (as from failure of Roosa pump seals or HEUI injector o-rings) can be detected by the loss of viscosity and the telltale odor.

Oil leaks

In rough order of frequency, the most common sites of oil leaks are:

- Valve cover gasket
- Oil pan gasket
- Timing case gasket
- Front and rear crankshaft seals
- Oil cooler
- Camshaft and oil gallery plugs

Tracing the source of a small leak can be difficult, because oil tends to migrate down and back, toward the rear of the engine. A black light detector or the aerosol powders sold for this purpose help, but the ultimate tactic is to pressurize the crankcase. Connect a low-pressure (5-psi maximum) air line to the dipstick tube. Soap the engine down and look for bubbles. Light foaming at valve cover gaskets is permissible. The rope-type crankshaft seals used in older engines might leak air and still function when the engine runs; lip seals may foam, but should not hemorrhage.

Internal oil leaks past bearings shows up as a loss of pressure, a condition that should be verified with an accurate test gauge made up to the main oil gallery (Figure 4-13). A pressure drop when loads are applied to a warm engine indicates excessive bearing clearances.

Oil enters the coolant by way of leaks in oil cooler or, more rarely, head gasket. Pressure test the cooler as described in Chap. 8, "Lubrication systems"; the head must be removed to detect leaks across the gasket, which makes the exercise a bit redundant.

The lower, external oil seal on HEUI injectors is all that stands between high-pressure lube oil and the fuel supply. A leak here can empty the sump within a few miles of driving. See Chap. 6 for more information.

4-13 Verify low oil pressure with an accurate gauge connected to a test port on the main gallery, as shown here on a Peugeot engine.

Coolant leaks

External leaks are usually visible; identifying internal coolant leak paths requires some detective work. Massive leaks into the combustion chamber produce white smoke at normal engine temperatures. Such leaks can sometimes be seen as coolant flow through the EGR valve.

A burst of steam when the dipstick is touched to a hot exhaust manifold signals the presence of water in the sump. Ethylene glycol coats the stick with brown varnish almost impossible to wipe off. Have the oil tested and, if the test shows the presence of antifreeze, prepare to tear the engine down for cleaning. Large accumulations of coolant in the sump convert the oil to a white, mayonnaise-like emulsion.

Excessive oil consumption

Blue smoke is the primary indicator of oil burning, although catalytic converters can disguise the condition. The question facing the mechanic is the source of the oil. Blue smoke at all speeds, hot or cold, points to cylinder/ring wear, which can be verified with tests described below. Smoke during acceleration after periods of idle and immediately upon startup indicates leaking valve seals. Oil in the intake manifold and/or air boxes originates from an upstream source, usually leaking turbo or blower seals. But other possibilities should not be overlooked. Sometimes merely changing the air filter element corrects the problem.

Loss of compression

Disabling one injector at a time will isolate any weak cylinders. If changing out the associated injectors does not restore power, the problem is almost surely compression-related. Loss of compression across all cylinders will be accompanied by high oil consumption and a noticeable reduction in power output.

The cylinder compression gauge (Figure 4-14) is an essential tool for diesel mechanics. Test procedures vary with make (e.g., some manufacturers specify a cold reading, while others require that the engine be at normal operating temperature). But the general procedure is as follows:

1. Cut off the fuel supply or disconnect the camshaft position sensor on computerized engines. Otherwise, the cylinder under test can fire and destroy the gauge.
2. Disable all cylinders by (depending upon how the gauge is mounted) removing the glow plugs or injectors.
3. Test each cylinder in turn, allowing about six compression strokes per gauge reading.
4. Check the readings against published specifications and against each other. Variations of 20% or more between cylinders results in noticeable roughness.

WARNING: do not squirt oil into the cylinder in an attempt to determine if the compression loss is due to rings or valves. The cylinder might fire.

4-14 Ford-supplied cylinder compression gauge.

Figure 4-15 illustrates a GM leakdown test kit with glow-plug adapter. Unlike competitive testers, the unit does not measure leakdown percentage, but it does indicate the source of air leaks, which is the purpose of the exercise.

1. Remove the radiator cap and warm the engine to operating temperature.
2. Shut off the engine and, using the timing marks, position No. 1 cylinder at top dead center (tdc). Finding tdc for the remaining cylinders is a bit challenging, but can be accomplished with a degree wheel or with the help of the whistle supplied with the kit.
   WARNING: if a piston is not at or very close to tdc, the engine will "motor," catching a hapless mechanic in the fan or belts.
3. Install the tester and apply about 75 psi of compressed air to it.
4. The significance of the various leak sites is listed in Figure 4-16.

4-15 GM cylinder leakdown tester includes air-line adapters, regulator, gauge, and top dead center (tdc) whistle. The whistle sounds as the piston nears tdc on the compression stroke.

Dipstick or oil fill: worn or sticking rings

Bubbles in coolant: warped or cracked cylinder head or block, blown head gasket

Air inlet: bad intake valve

Leaking from glow plug hole on adjacent cylinder: blown head gasket

Exhaust: bad exhaust valve

4-16 Potential leak sites for compression.

The amount of blowby past the rings is a rough indicator of cylinder bore/ring wear. Figure 4-17 illustrates the test apparatus, consisting of a pressure gauge and an adaptor that replaces the oil filler cap. PCV atmospheric vents must be sealed. A major limitation of this test is that few manufacturers provide rpm/blowby pressure specifications. The mechanic must establish the norm while the engine is still healthy.

Oil analysis

Short of tearing the engine down for inspection, oil analysis is the best index of wear. The technology, developed initially for diesel locomotives and perfected by the

4-17 A crankcase pressure tester is part of the Ford PN 0190002 gauge bar set.

U.S. Navy, can predict engine failure with some exactitude. The oil sample undergoes spectroscopic analysis to identify 16 or more elements with an accuracy of within one part per million. Bearing wear shows up as lead, silver, tin, and aluminum; liner and ring wear as iron and chromium; valve wear as nickel; and bushings as copper. Silicon and aluminum particles suggest air filter failure.

5
CHAPTER

Mechanical fuel systems

Diesel fuel systems were entirely mechanical until the mid-1980s, when manufacturers of vehicle engines turned to computers to give more precise control over fuel delivery and timing. Mechanical systems continue to be specified for some marine engines and for large, mostly low-speed industrial power plants.

Air blast

Early engines used compressed air to force fuel into the cylinders against compression pressure. Injectors received fuel from a low-pressure pump and air from a common manifold, which was pressurized to between 800 and 1000 psi. The engine camshaft opened the injectors, appropriately enough called "valves," to admit a blast of compressed air and atomized fuel to the cylinders. A throttling valve regulated engine speed by controlling the amount of fuel delivered; air required no metering, since surplus air is always present in diesel cylinders.

Air injection had serious drawbacks. The air/fuel mixture could not penetrate deeply into cylinders that were themselves under 450–500 psi of compression pressure. Attempts to increase engine power by admitting more fuel merely dampened the flame. The compressors were parasitic loads that absorbed 15% and sometimes more of engine output. Nor did air injection make for a compact package—starting and four-stage air-injection compressors accounted for a third of the length 1914-era Krupp submarine engines.[1]

Early common-rail

The need for more compact and silent submarine engines led the British arms maker Vickers to develop airless, or solid, injection. One can make the case that the ultimate consumers of nineteenth- and twentieth-century technology were enemy soldiers,

---

[1] C.L. Cummins, Jr., "Diesel Engines of 2000 BHP per Cylinder—Pre-1914 Marine Engine Developments," *History of the Internal Combustion Engines,* ed., E.F.C. Somerscales and A. A. Zagotta, p. 19, 1989.

5-1 The common, or third-rail system supplied fuel to cam-operated injectors from a header, pressurized by a remote pump. By the late 1920s, c-r was being phased out by Bosch inline pumps, only to be revived again as the century closed. These new common-rail systems are discussed in the following chapter.

sailors, and civilians. At any rate, solid injection was a vast improvement and, after the end of hostilities in 1919, was widely commercialized. Injectors fed from a common rail that was filled with high-pressure fuel and usually fitted with an accumulator to smooth pressure variations generated by injector opening and closing (Figure 5-1). Adjustable wedges between the camshaft lobes and the injector followers permitted injection duration to vary with speed and load.

Even so, this early form of common-rail injection limited engine speed, since increasing the effective width of the wedge by forcing it deeper into contact with the cam also advanced the timing, causing the injector to open earlier. There were also maintenance concerns. According to contemporary accounts, the rails leaked and injectors dribbled fuel throughout the whole operating cycle.

Jerk pump system

The big breakthrough came with the development of the Bosch inline pump, used today in a form almost identical to the first production run of April, 1927. The first pumps were mated with pintle injectors for indirect injection (IDI) applications. Direct injection (DI) multiple-orifice injectors arrived in 1929 and, two years later, Bosch integrated centrifugal governors with the pumps. At this point, the basic pattern for modern diesel engines was set. Electronic engine management systems and exhaust emission controls are add-ons.

The inline pump consists of a row of individual plungers, one per engine cylinder, operated by the same internal camshaft. High-pressure tubing connects each of the plungers to its injector. The spring-loaded injectors function like pop-off valves to open automatically when a certain pressure threshold is reached. The sudden loss of line pressure gave rise to the term "jerk pump," which while a bit inelegant, is descriptive.

The distributor pump was developed in the early 1960s as a means of reducing the number of extremely precise and expensive plungers. One pumping unit, consisting of one, two, or sometimes three plungers, serves all injectors. After pressurization, the fuel passes through a rotary valve, known as a distributor head, for allocation to the individual injectors. A distributor pump is the hydraulic equivalent of an ignition distributor. Because of their relatively low cost, distributor pumps are often used on automobiles and light trucks. The major disadvantage is that the internal cam, which drives the plunger set, depends solely upon fuel for lubrication. Inline pumps lubricate the high-pressure cam faces with motor oil, either from an internal reservoir or from the engine oiling system.

Figure 5-2 illustrates the layout of the jerk-pump system that, in this example, employs an inline pump. A distributor-type pump could be substituted.

A lift pump, shown on the lower left of the drawing, delivers fuel to the filters and from there to the suction side of the inline pump. High-pressure fuel exits the pump through dedicated lines to each injector. A second line made up to each injector recycles surplus fuel back to the tank. The system works at three pressure levels—low pressure, on the order of 30–50 psi, between the lift pump and injector pump, pressure of several thousand psi in the injector piping, and slightly more than zero pressure on the return line.

Inline pumps

The parts breakdown for a Bosch size A Series PE inline pump for six-cylinder engines is shown in Figure 5-3. Shims (21 and 49) establish camshaft float. Two fuel-delivery adjustments are provided. The basic setting is determined by the height of

A Primary fuel filter and water separator
B Secondary fuel filter
C Excess fuel return tube
D Injection pump overflow tube

5-2 An inline-pump system as used on Ford-based Lehman marine engines.

5-3 Bosch Series PE pump. This example is found on six-cylinder Chrysler marine engines.

1 Delivery valve holder	21 Adjusting shim	41 Roller bushing
2 Lockplate	22 Tapered roller bearing	42 Screw plug
3 Lockwasher	23 Nut	43 Camshaft
4 Bolt	24 Lockwasher	44 Key
5 Delivery valve spring	25 Stud bolt	45 Lockwasher
6 Delivery valve gasket	26 Screw plug	46 Screw
7 Delivery valve assembly	27 Gasket	47 Center bearing
8 Plunger assembly	28 Oil level gauge	48 Distance ring
9 Air bleeder screw	29 Pinion clamp screw	49 Adjusting shim
10 Gasket	30 Control pinion	50 Tapered roller bearing
11 Air bleeder plug	31 Control sleeve	51 Oil seal
12 Gasket	32 Upper spring seat	52 Bearing cover
13 Pump housing assembly	33 Plunger spring	53 Lockwasher
14 Air bleeder assembly	34 Lower spring seat	54 Screw
15 Gasket	35 Adjusting bolt	55 Control rack
16 Plate cover	36 Locknut	56 Adapter
17 O-ring	37 Tappet	57 Connector ring
18 Gasket	38 Tappet guide	58 Gasket
19 Setscrew	39 Roller pin	59 Connector bolt
20 Distance ring	40 Roller	

the adjustment bolts (35) that thread into the tops of each tappet. In addition, friction clamps (29) permit the control-sleeve pinions (30) to move relative to the rack to compensate for tooth wear and production tolerances. Fuel-delivery adjustments for this and all other injection pumps must be made on a test stand.

The drawing of a single-plunger pump in Figure 5-4 helps to clarify the relationship between the rack, fuel control gear (plunger pinion), and plunger. Fuel enters through the intake port on the right and exits past the delivery valve at the top of the unit. The obliquely cut groove on the plunger outer diameter (OD) functions in conjunction with the rack to throttle fuel delivery.

Delivery valve (5)

Intake hole (4)

(1) Plunger

(3) Obliquely cut lead

(2) Plunger barrel

Fuel injection volume increases

Fuel injection volume decreases

(8) Fuel control rack

(6) Fuel control gear

(7) Flange

5-4 Plunger, barrel, and rack assembly. The relief labeled "obliquely-cut lead" is generally called the "helix."

How this is done is shown in Figure 5-5. At the bottom of the stroke, the plunger uncovers the inlet port. Fuel enters the pressure chamber above the plunger. The plunger rises, initially pushing fuel back out the inlet port. Further movement masks the inlet port. The plunger continues to rise, building pressure on fuel trapped above it. The delivery valve opens and a few milliseconds later the injector discharges. Fuel continues to flow until the annular groove milled along the side of the plunger uncovers the inlet port. At this point, pressure bleeds back through the inlet port and injection ceases. Because of the shape of the groove, rotating the plunger opens the inlet port to pressure earlier or later in the plunger stroke.

American Bosch, Robert Bosch, and CAV barrels are drilled with a second port above the inlet port to accept spillage during part-throttle operation. Figure 5-6 illustrates the metering action of a CAV pump. Fuel enters the barrel at A and continues to flow until plunger movement masks the two ports (inlet shown on the left, spill port on the right). At full load, the pressure bleed-down through the

No injection   Half injection   Full injection

Delivery chamber

Plunger

Oil hole

Lead groove

Effective stroke

Effective stroke

5-5 Constant-beginning, variable-ending plunger action with surplus fuel exiting through the inlet port.

spill port is delayed until the plunger approaches the end of its stroke, as shown in drawing C. You might wonder why the spill port opens before top dead center (tdc), when a few more degrees of cam movement would raise fuel pressure even more. The reason is that cam-driven plungers behave like pistons, accelerating at mid-stroke and slowing as the dead centers are approached. Opening the spill port early, while plunger velocity and pressure rise are rapid, terminates injection far more abruptly than if the port remained closed until the plunger reached tdc.

In drawing D, the annular groove has rotated to the half-load position. The effects of further rotation are shown at E, which represents idle, and at F, shutdown.

However the porting is arranged, the effective stroke has a constant beginning and a variable ending for pumps this book is concerned with. Some marine and large stationary engines meter fuel delivery at the beginning of the stroke.

Details of the mechanism for transmitting rack movement to the plungers vary with the manufacturer, but always include an adjustment to equalize delivery between plunger assemblies. At a given rack position, each cylinder must receive the same amount of fuel.

A   B   C   D   E   F

5-6 A variation of the principle illustrated in Figure 5-5, using a spill port to vent the unneeded fuel. GM Bedford Diesel

Distributor pumps

The main technical advantage of distributor pumps is that each cylinder receives precisely the same amount of fuel since all cylinders feed from the same plunger or plunger set. And, as pointed out above, reducing the number of plungers, which must be lapped to angstrom tolerances, reduces costs.

Lucas/CAV

Figure 5-7 illustrates a CAV system for medium trucks. The engine-driven feed pump moves fuel from the tank to the filter and regulating valve. Fuel then passes to the transfer pump and, from there, to the inlet side of the distributor pump for pressurization and injection. As explained below, the regulating valve throttles the injection pump during low-speed operation. Surplus fuel returns to the tank through low-pressure lines.

A pair of cam-driven plungers, mounted on the pump rotor and turning with it, generate injector pressure (Figure 5-8). During the inlet stroke the plungers move outward under pressure from fuel entering through the inlet (or metering) port. As the rotor continues to turn, the inlet port is blanked off by the rotor body and one of the outlet ports is uncovered.

As the outlet port is uncovered, the internal cam, acting through rollers, forces the plungers together. Fuel passes through the port for injection. Fuel that slips by the plungers returns to the filter.

The amount of fuel delivered per plunger stroke is determined by the regulating valve (upper right in Figure 5-7) that is controlled by the accelerator pedal and the

5-7 Schematic of a Lucas-CAV distributor-pump system. GM Bedford Diesel

5-8 Opposed-piston distributor pump operation. GM Bedford Diesel

centrifugal governor. At low engine speeds, the regulating valve limits the pressure of fuel going to the transfer pump and into the injection pump. Because the pump plungers are driven apart by incoming fuel, their outward displacement is determined by transfer-pump pressure. As engine speed and/or load increase, the regulating valve increases fuel pressure to force the pump plungers further apart. Unlike Bosch inline pumps, which throttle by varying the effective stroke, the Lucas/CAV unit varies the actual stroke.

Bosch VE

Nearly 50 million VE pumps have been produced since introduction in 1975 with applications ranging from fishing boats to luxury automobiles. Figure 5-9 sketches a basic installation, with filter, gravity-type water separator (sometimes called a sedimentor), fuel lines from the pump and injectors, and a solenoid-operated fuel shutoff (magnet) valve. Normally the fuel shutoff requires battery power to open; for marine applications the valve must be energized to close. This permits the engine to continue to run should the vessel lose electrical power.

The VE turns at half engine speed and is geared to a mechanical (shown in Figure 5-10) or electronic governor. The rear half of the pump houses the vane-type transfer pump and regulator that supplies the high-pressure section with fuel at pressures ranging from 40 psi at idle to 175 psi at full throttle. Vane-pump pressure also controls the hydraulic timer that advances injection with increased engine speeds.

A tongue-and-groove joint mates the rear half of the two-piece drive shaft with the forward, or plunger, half. This joint, which can be seen clearly in Figure 5-11, permits the forward half of the shaft to move fore and aft as it rotates under the impetus of the face cam. The cam reacts against the housing through roller bearings, oriented as shown in Figure 5-12.

A control sleeve regulates the effective stroke of the plunger, which discharges through delivery valves on the distributor head.

5-9 Typical small-engine distributor-type system. Yanmar Diesel

5-10 Bosch VE Series injection pump.

5-11 Cross-sectional view of the VE. Notice the integration and compactness of the design.

5-12 The VE plunger reciprocates in the pumping function and rotates in the distributor, or fuel-allocation, function. Yanmar Diesel

1 Plug
2 Gasket
3 Valve holder
4 Adjusting screw
5 Spring
6 Ball valve

5-13 Fuel overflow valve functions as an automatic bleeder, diverting air and surplus fuel back to tank. Marine Engine Div., Chrysler Corp.

Injector pump service

Bleeding

Because surplus fuel recycles to the tank or filter through fuel overflow valves mounted at high points on pumps and injectors, a half-hour or so of operation is generally sufficient to purge the system of air bubbles (Figure 5-13). More serious air intrusions require bleeding, using procedures described in Chap. 4 for conventional, pre-computer engines and elaborated upon in Chap. 6 for more modern engines.

Timing

Injection pumps have two provisions for synchronizing fuel delivery with piston movement. Timing marks on the drive gears establish the basic relationship. Elongated mounting-bolt holes, which allow the pump body to be rotated a few degrees, provide the fine adjustment. Reference marks stamped on the pump body and mounting flange enable the adjustment to be replicated (Figure 5-14).

5-14 Timing marks on a Bosch distributor pump. Ford Motor Co.

5-15 Transparent section (4) of fuel pipe adapter indicates plunger movement for timing purposes. Lombardini

Before removing a pump, bar the engine over until both valves on No. 1 cylinder close and the timing mark on the harmonic balancer or flywheel aligns with its pointer. This procedure indexes pump-gear timing marks for easy assembly. If the same pump is reinstalled, the reference marks on the pump body and flange should be valid. Substituting another pump puts the marks into question and the engine should be retimed, either statically or dynamically.

Static timing procedures vary enormously, but the purpose of the exercise is to synchronize the onset of fuel delivery with the piston in No. 1 cylinder. Depending upon engine make, model, and application, fuel should begin to flow anywhere from 8–22° btdc as the flywheel is barred over by hand.

Flywheels for small utility engines generally have two marks inscribed on their rims—one representing tdc and the other, always in advance of the first, indicating when fuel should begin to flow from No. 1 delivery valve. A convenient way to monitor fuel flow is to make up an adapter out of a length of clear plastic tubing and a delivery-valve fitting, as shown in Figure 5-15. The mechanic slowly bars the engine over, while watching for the slightest rise in the fuel level. The onset of fuel movement should occur at the moment the timing mark aligns with its pointer. If tdc is passed and the plunger retreats, the flywheel must be turned back 15° or so to absorb gear lash, and the operation repeated.

Drive gears for Navistar (International) DT358 and its cousins have six timing marks. Which one to use depends upon the engine model and application. In a reversal of traditional practice, certain American Bosch pumps time to the end, rather than the onset, of fuel delivery. The delivery valve, which acts as a check valve, for No. 1 plunger must be disabled before timing.

Timing specifications for distributor pumps are often expressed in thousandths of an inch of plunger movement from bdc. Figure 5-16 illustrates the dial-indicator adapter that replaces the central bolt in the distributor head. Locating

5-16 For critical applications, plunger travel is measured with a dial indicator. Ford Motor Co.

bdc—the precise moment when the plunger pauses at the bottom of its stroke—requires patience. Once bdc is found, the mechanic zeros the gauge and bars the engine over in the normal direction of rotation to the appropriate crankshaft or harmonic-balancer mark. He then rotates the pump body as necessary to match lift with the published specification.

Dynamic timing, made with a strobe light while the engine ticks over at slow idle, compensates for pump-gear wear and other variables. It is the only way that cargo van and other inaccessible engines can be timed.

The Sun timing light draws power from a wall outlet or, if equipped with an inverter, from the engine's 12-V or 24-V batteries (Figure 5-17). A transducer clamps over No. 1 fuel line to trigger the strobe when the injector opens and the sudden drop in fuel pressure contracts the line. The instrument also tracks how many crankshaft degrees the timing advances as engine speed increases.

5-17 Sun-timing light triggers from pulses in the fuel pipe.

Unit injectors (UIs), which integrate the pump function with injection, are timed as described below:

Repair

Other than periodically checking the oil level for in-line pump reservoirs and adding rust inhibitor to the fuel before long-term storage, injector pumps require no special attention. When loss of pressure indicates that the pump has failed, the mechanic farms the unit out to a specialist for cleaning or repair.

While most mechanics would not tackle a Bosch VE, inline pumps are simple devices that can be opened for cleaning and elementary repairs. But the work requires patience and the highest possible levels of cleanliness. Plungers are lapped to their barrels and must be assembled as found. Nor should these sensitive parts be touched with bare fingers. Wear surgical gloves or use forceps. Tappets and other adjustable parts must be kept with their plungers. When disassembly entails loss of adjustment, the existing adjustments must be scribe marked. If at all possible, obtain a drawing of the pump before you begin.

Inline pumps come apart as follows:

- Unscrew the delivery valves mounted above each barrel and lay them out in sequence on the bench, which should be covered with newspaper to reduce the possibility of contamination.
- Most inline pumps incorporate a side cover for tappet access; others mount the individual plunger assemblies from above with studs. On those with side covers, rotate the camshaft to bring each tappet to the top of its stroke and shim the tappet. Withdraw the plungers from the top with the aid of a hooked wire or expansion forceps. Mark or otherwise identify the plungers for correct assembly.
- Inspect the lapped surfaces with a magnifying glass. Deep scores or pronounced wear marks mean that the pump has come to the end of its useful life. Overtightening the delivery valves warps the barrels and produces uneven wear patterns. Water or algae in the fuel leave a dull, satiny finish on the rubbing surfaces. Loss of the sharp edges on the helix profile upsets calibration.
- A plunger should fall of its own weight when the barrel is held 45° from vertical. Gummed or varnished plungers can make governor action erratic and accelerate wear on the helixes and rack.
- If you have the specifications, check the cam profile with an accurate (i.e., a recently recalibrated) micrometer.
- Examine the rack teeth for wear. Figure 5-18 illustrates, in exaggerated fashion, the wear pattern. The rack should move on its bushings with almost no perceptible side or vertical play. Installation of new bushings requires a factory reamer.

Delivery valves

Injectors feed through delivery valves mounted on the distributor head or above each barrel on inline pumps. These check valves open under pressure to supply fuel to the injectors and close automatically during the suction stroke. Most delivery valves consist of a conical sealing element shaped like an inverted top hat (Figure 5-19). The extension below

New

Worn-knife edge

5-18 Control-rack teeth bell-mouth in service.
GM Bedford Diesel

the element, known as the piston, has two functions. It centers the valve over its seat and acts as a ram to force fuel back into the pump as the valve closes. This action assures a rapid drop in pressure so that the injector snaps shut without after-dribble.

Delivery valves come in various shapes and sizes. Some replace the conical element with a disc, others hold a certain percentage of pressure in the downstream plumbing after closing. But all work on the same general principle.

Service

Delivery-valve problems are cylinder specific. If the valve sticks open, no fuel passes to the associated injector. Leakage is harder to diagnose. White smoke that persists after all bases have been touched—pump pressure and timing, new injectors, and engine compression—suggests that one or more delivery valves may be at fault. Disassemble one valve at a time, clean the parts thoroughly, and test for leaks by blowing through

Closed   Open

5-19 Fuel delivery valve operation. GM Bedford Diesel

5-20 Delivery valve test. Check valve shown at 1, piston at 2.
Marine Engine Div, Chrysler Corp

the outlet port. Older, simpler valves can sometimes be resurfaced. Check piston fit by depressing the valve and placing your finger over the inlet port (Figure 5-20). You should feel the vacuum as the piston falls.

Injectors

Mechanical injectors consist of two main parts—an injector body, or nozzle holder, and a needle (Figure 5-21). The injector body secures the assembly to the cylinder head and includes connections for fuel inlet and return lines.

A spring-loaded needle ("nozzle valve") in the drawing, controls fuel flow through the injector. At NOP fuel pressure exerts sufficient force on the raised shoulder near the tip of the needle to overcome spring tension. The needle retracts clear of the orifice and injection commences. Once the needle lifts, its full cross-sectional area—the shoulder and tip—comes under fuel pressure. Consequently, less pressure is required to hold the needle open than to unseat it. This feature holds the nozzle open as fuel pressure drops in response to injection. When the pump plunger descends and the delivery valve closes, fuel pressure drops abruptly. The return spring forces the needle down to seal the orifice and injection ceases.

The smaller the nozzle orifice, the finer the fuel spray and, all things equal, the less ignition lag. But if too small, the orifice limits engine output by restricting fuel flow. Multi-hole nozzles, some with a dozen or more, electrically discharge machined orifices that satisfy both flow rate and atomization requirements, have become standard on DI engines.

IDI engines employ pintle nozzles, two varieties of which are illustrated in Figure 5-22. The pintle is an extension of the needle that, when retracted, produces a hollow, cone-shaped spray. Because of the large diameter of the orifice and the shuttle action of the pintle, these nozzles are less susceptible to carbon buildup than multi-hole types.

The throttling pintle introduces fuel in a staged sequence to reduce the sudden rise in pressure and heat during initial combustion. The pintle has a reduced diameter

1. Leak off union
2. Body cap nut
3. Seal washers
4. Fuel inlet passage
5. Nozzle cap nut
6. Nozzle
7. Nozzle valve
8. Nozzle valve spindle
9. Injector body
10. Spring cap locknut
11. Nozzle valve spring
12. Spring seat washer
13. Spring cap nut

5-21 Cross-section and exploded view of two CAV injectors. GM Bedford Diesel and Lehman Ford Diesel.

Pintle nozzle   Yanmar semi-throttle nozzle

5-22 Pintle (left) and throttling pintle nozzles. Yanmar Diesel

on the tip that, when seated, extends into the combustion chamber. Yanmar and most other manufacturers use a throttling pintle with the tip in the form of an inverted cone, shown on the right in Figure 5-22. During the initial phase of injection, the cone lifts so that its base almost fills the orifice. At full lift the cone retracts clear of the orifice for maximum fuel delivery.

Injector service

Regardless of the nozzle type, injectors should:

- Discharge cleanly without "after-dribble," which tends to collect in the cylinder between cycles.
- Atomize fuel for rapid combustion.
- Direct the spray into the far reaches of the cylinder, but without impinging against the piston or chamber walls.

Most malfunctions are caused by carbon buildup on the tip, which distorts the spray pattern and can sometimes stick the needle or pintle in the open position. Pressure-activated injectors can be tested with the apparatus shown in Figure 5-23. Similar tests can be performed on unit injectors by loading the pump cavity with fuel and applying force to the pump plunger.

WARNING: Wear eye protection and heavy gloves when testing injectors. Fuel spray can cause blindness and blood poisoning.

There are four standard tests—spray pattern, NOP, sealing effectiveness, and chatter.

Spray pattern

Connect the tester as shown and, operating the pump lever in short, rapid strokes, observe the spray pattern. Pintle nozzles should produce a finely divided spray of uniform consistency and penetration (Figure 5-24). Multi-hole nozzles generate a wide, fan-like pattern (Figure 5-25). It can be helpful to direct the discharge against a piece of paper and compare the pattern to that produced by a new injector.

5-23 Ford-supplied injector tester. These tools have multiple uses, including testing common-rail fuel manifolds and HEUI oil galleries for leaks.

Nozzle opening pressure

Slowly depress the pump lever and note the peak gauge reading. Compare this pressure with the manufacturer's NOP specification. Some specifications are fairly broad, and engines may not idle properly unless NOPs cluster at one or the other end of the specification range. Most needle-return springs can be stiffened with shims.

Sealing

Dribble standards have become tighter in recent years. The pintle injector illustrated in Figure 5-26 should remain fuel tight under 80% of NOP for at least 10 seconds. Note and correct any other source of leakage.

Chatter

Injector opening is normally accompanied by a sharp "pop," which suggests that the needle retracts cleanly. Verify this action as the pump lever is cycled rapidly. However, leaking injectors can usually be made to chatter.

Timers

Moving fuel through an open injector requires time and, once injected, additional time is needed for the fuel to ignite. As engine speeds increase, less time is available to accommodate these delays, and the start of injection must occur earlier.

The drawing back at Figure 5-10 shows the VE hydraulic timer in cross-section. Pressure developed by the feed pump rises in an almost linear fashion with engine speed. As pressure increases, the piston moves to the right, compressing its return spring.

Stream
- Injection pressure low
- Nozzle seized
- Nozzle spring broken
- Dirt on valve seat

Spike
- Injection port damaged or dirty
- Carbon build-up
- Nozzle end abnormally worn

Spray
- Injection port worn
- Carbon build-up

Slanted
- Uneven seat contact
- Injection port damaged or worn
- Carbon build-up

Normal

5-24 Spray pattern diagnosis—modified pintle nozzle, IDI engines.

Correct

Spray from each nozzle
hole is uniform

Poor

- Excessive difference in spray angle (ϑ)
- Excessive difference in injection angle (a)
- Incomplete atomization
- Sluggish starting/stopping of injection

5-25 Spray pattern diagnosis—multi-hole nozzle, DI engines.

A roller transfers this motion to the drive cam, turning it few degrees against direction of shaft rotation.

Mechanical timers are similar to SI advance mechanisms, in that they employ centrifugal weights to sense engine speed and advance the cam (Figure 5-27). Spring tension on the weights determines the cut-in speed and shims fix the travel limit.

To check timer operation, connect a timing light to an injector pipe and run the engine up to speed. Injection should advance smoothly without hesitation as the rack is extended. Specifications vary, but most applications require 5–7.5° of total advance.

Diaphragm controls

Injector pumps often include a diaphragm-operated control that reduces fuel delivery at high altitudes and/or increases fuel at wide throttle angles for turbocharged applications. Diaphragms are tested with a vacuum gauge.

Acceptable No signs of any fuel	**1**
Acceptable No visible fuel but damp	**2**
Not acceptable Visible fuel and wet	**3**
Not acceptable Drop forms but does not fall or run along bottom tip	**4**
Not acceptable Drop falls or runs along bottom of the tip	**5**

5-26 Dribble test—pintle nozzle, IDI engines.

Centrifugal governors

Spark-ignition engines have a built-in governor in the form of a throttle valve that, even wide open, limits the amount of air passing into the engine. Diesels have no such limitation, since surplus air is always available for combustion. Engine speed depends solely upon the amount of fuel delivered. If the same volume of fuel necessary to cope with severe loads were delivered under no-load conditions, the engine would rev itself to destruction. Consequently, all diesel engines need some sort of speed-limiting governor.

Most governors also control idle speed, a task that verges on the impossible for human operators. This is because the miniscule amount of fuel injected during idle exaggerates the effects of rack movement. Automotive applications are particularly critical in this regard. Idle speed must be adjusted for sudden loads, as when the air-conditioner compressor cycles on, and when the driver turns the wheel and engages the power-steering pump.

In addition to limiting no-load speed and regulating idle speed, many governors function over the whole rpm band. The operator sets the throttle to the desired speed and the governor adjusts fuel delivery to maintain that speed under load. The degree of speed stability varies with the application. No governor acts instantaneously—the engine slows under load before the governor can react. Course regulation holds speed changes to about 5% over and under the desired rpm, and is adequate for most applications. Fine regulation, of the kind demanded by AC generators, cuts the speed variation by half or less.

Centrifugal governors sense engine speed with flyweights and throttle position with spring tension. Figure 5-28 illustrates the principle—as engine speed increases,

1 Timer flange
2 Timer spring
3 Adjusting shim
4 Flyweight
5 Flyweight-holder
6 Nut
7 Lock washer
8 Lock plate
9 Adjusting shim
10 Thrust washer
11 Timing gear
12 Bolt
13 Lock washer
14 Round nut
15 Lock washer

5-27 Centrifugal fuel timer varies injection timing in response to changes in engine speed.

the spinning flyweights open to reduce fuel delivery. As the throttle is opened, the spring applies a restraining force on the flyweight mechanism to raise engine speed.

So much for theory. In practice, centrifugal governors demonstrate a level of mechanical complexity that seems almost bizarre, in this era of digital electronics. But millions of these governors are in use and some discussion of repair seems appropriate.

Service

Governor malfunctions—hunting, sticking, refusal to hold adjustments—can usually be traced to binding pivots. In some cases, removing the cover and giving the internal parts a thorough cleaning is all that's necessary. If more work is needed, the governor and pump should be put into the hands of a specialist.

Bosch type

5-28 Bosch-type centrifugal governor. The flyweights react against throttle-spring tension to set the speed of the engine. As loads are encountered, the engine slows, the flyweights exert less force, and the spring extends the rack to restore engine rpm. Yanmar Diesel Engine Co., Ltd.

To appreciate why this is so, consider the matter of bushing replacement, which seems like a simple enough job. Using the CAV unit as a model, the critical bushings are those at the flywheel pivots (225 in Figure 5-29) and the sliding bushing (231) on the governor sleeve. To remove and install the flyweights you will need CAV special tool 7044-8. Bushing press in, but factory reamers are needed to size them properly. And a governor with enough hours on it to wear out bushings will almost surely need

5-29 CAV governor in exploded view.

recalibration, which can only be accomplished with a factory tooling that correlates piston stroke with rack position.

Pneumatic governors

Pneumatic, or flap valve, governors are still sometimes encountered. The best that can be said for these devices is that they are easier to repair than centrifugal or electronic governors. The velocity of air moving through the intake manifold is a function of piston speed. The faster the engine runs, the greater the velocity. Pneumatic governors sense this velocity as vacuum developed by a venturi mounted on the air intake.

The venturi restricts the flow of air and, so doing, reduces its pressure. This relationship has a constant of 20, as shown in Figure 5-30. If we use consistent terms and multiply pressure times velocity at any point in the venturi, the answer is always 20, or something approximating that number. Nor need the venturi be the sort of streamlined restriction shown in the drawing; any impediment accelerates air flow and reduces its pressure.

MZ

This unit, shown in cross-section in Figure 5-31, uses a flap, or butterfly, valve (4) as a variable venturi. A tube bleeds vacuum from the edge of the valve to the left or low-pressure side of the diaphragm housing. A second tube brings filtered air at atmospheric pressure to the right side of the housing. The spring-loaded diaphragm (8) separates these two halves of the housing and connects to the fuel rack.

The flap valve is free to pivot in response to incoming air velocity. At low engine speeds the valve positions itself as shown in the drawing. All air entering the engine must accelerate and squeeze past the obstruction created by the edge of the valve and the vacuum-line connection. The diaphragm responds to the resulting depression by moving toward the left, or low-pressure, side of the housing. Fuel delivery is reduced. If the engine decelerates as when encountering a load, vacuum drops and the diaphragm shifts toward the right to increase fuel flow.

At wide-open throttle, the velocity of incoming air pivots the flap valve full open. Sensing the low level of vacuum, the diaphragm shifts to the right, extending the rack.

V = 5
P = 4
C = 20

V = 10
P = 2
C = 20

V = 5
P = 4
C = 20

5-30 The venturi constant.

1 Air cleaner
2 Auxiliary venturi
4 Butterfly valve
5 Idler spring
6 Auxiliary idler spring
7 Mainspring
8 Diaphragm
9 Diaphragm housing
10 Adjusting shim
11 Balance (Angleich) spring
12 Pushrod
13 Adjusting shim
14 Cotter pin
15 Washer
16 Connecting bolt
17 Nut
18 Tubing connection
19 Control rack
20 Lever
21 Washer
22 Screw
23 Stop lever
24 Stroke setscrew
25 Stop bolt
26 Governor housing
27 Bolt
28 Gasket
29 Adjusting shim
30 Endplate
31 Oil seal
32 Cap
33 Bolt
34 Lockwasher
35 Split pin
36 Shaft

5-31 MZ centrifugal governor.

1 Diaphragm
2 Governor spring
3 Auxiliary idler spring
4 Stroke setscrew
5 Stop bolt
6 Stop lever
7 Control rack
8 Connecting bolt
9 Pushrod
10 Adjusting shim
11 Balance (Angleich) spring
12 Adjusting shim

5-32 MZ balance mechanism.

The governor also includes the mechanism shown in Figure 5-32 that reduces fuel delivery at idle. As the diaphragm moves to its extreme leftward position, it brings the stop lever (6) to bear against the stop bolt (5). This action compresses the balance spring (11) and displaces the rack (7) toward the no-fuel position. As engine speed increases, the diaphragm moves to the right. The balance spring mechanism moves with it, away from the stop lever. Once clear of the lever, the spring has nothing to react against and is out of the circuit.

Service

The leather diaphragm should hold 500 mm/H_2O of vacuum with a leak-down rate of no more than 2 mm/sec. It should be flexible enough to collapse of its own weight when held by the rim. Nissan dealers can supply the special diaphragm oil required to soften the leather.

A screw under the circular end cover reacts with the auxiliary idle spring (3) to set the idle speed. Tightening the screw moves the rack to the right to increase fuel delivery and idle rpm. But the effect extends to all engine speeds (Figure 5-33). Fuel delivery for best idle over-fuels the engine at high speed and, conversely, a setting that gives best fuel economy at speed, costs low-end power. The adjustment is a compromise.

Other adjustments require a test fixture available to dealer mechanics. For example, the stroke adjustment, shown at 4 in Figure 5-32, is made to factory specifications with the diaphragm chamber evacuated and the pump turning 500 rpm. To further complicate matters, balance-spring adjustment must be done by comparing its action with a known good unit.

Unit injection

Thus far, we have discussed three types of mechanical fuel systems—air blast, early common-rail, and jerk pump. The last of these systems, and in some ways the most attractive, is unit injection.

5-33 MZ injection curves.

Unit injection combines the pumping function with injection. That is, each UI incorporates its own high-pressure pump driven from the engine camshaft. Injection pressures, often in excess of 20,000 psi, are confined entirely within the injector bodies. The lines that connect the lift pump with the injectors see only moderate pressures, on the order of 50 psi. Consequently, UI systems are virtually leak-proof. Mechanics, at least the ones old enough to remember Detroit Diesel two-strokes, like the way UIs localize problems. When a UI fails, only the associated cylinder goes out. And because an engine will run with one or two bad injectors, it becomes its own test stand.

Figure 5-34 illustrates the standard Detroit Diesel fueling circuit. The restrictor draws off a portion of the incoming fuel to cool the injectors before returning to

5-34 Unit injectors combine a high-pressure pump and injector nozzle in the same assembly. Detroit Diesel

the tank. Two stages of filtering are provided, with the primary filter on the suction side of the fuel pump.

In addition to the parts found on more conventional injectors, a UI has a mushroom-shaped cam follower, return spring, pump plunger, fuel-supply chamber, and one or more check valves (Figure 5-35). The rack controls fuel delivery by rotating the plunger as described under "Inline pumps."

5-35 Cutaway of a Detroit unit injector.

Fuel enters the injector and passes to the supply chamber. As the engine cam forces the plunger down, fuel trapped in the supply chamber comes under increasing pressure. At NOP, the needle valve lifts and injection begins. The disc-shaped check valve is fail-safe. Should needle fail to seat, the check valve closes to prevent air from entering the fuel supply.

Other than changing out plunger-related parts, unit injectors are not normally considered field-repairable.

Low-pressure system

Regardless of the type of high-pressure system, most engines employ similar low-pressure systems.

Lift pumps

A fuel supply, feed, transfer, or lift (the terms are interchangeable) pump supplies low-pressure fuel to the suction side of the injector pump or unit injectors. Most lift pumps are driven mechanically, although in recent years, electric drive has become popular. Pressure varies with 50 psi as a ballpark figure.

Stanadyne, Lucas/CAV and Bosch distributor-type injector pumps employ an integral vane-type lift pump as pictured back in Figure 5-9 and subsequent drawings. The sliding vanes rotate in an eccentric housing with the outlet port on the periphery.

Gear-type lift pumps, similar to lube-oil pumps, develop pressure from the tooth mesh and usually incorporate a spring-loaded pressure-limiting valve.

Chrysler-Nissan SD22 and SD33 engines feed through a piston pump driven off the injector pump (Figure 5-36). The operation of the pump is slightly unorthodox, in that fuel is present on both sides of the piston to eliminate air locks. As the piston retracts, fuel under it is expelled and the inlet check valve opens to admit fuel into the chamber above the piston (Figure 5-37). As the piston rises, the check valve closes to permit the upper chamber to be pressurized. Fuel under the piston then reverses course to fill the void left by the rising piston. Near tdc the spring-loaded discharge-side valve opens. Spring tension determines pump pressure. The upper chamber can also be pressurized manually to bleed the system.

Small engines are often fitted with low-pressure diaphragm pumps such as the AC unit pictured in Figure 5-38. The pump consists of a spring-loaded diaphragm activated by the engine camshaft. Inlet- and discharge-side check valves are often interchangeable.

Lift-pump service

Loss of pressure is the definitive symptom of pump failure, although it would be helpful if manufacturers provided output volume specifications. Before condemning the pump, check for air leaks as described in the previous chapter. A length of transparent plastic tubing spiced into the pump output line will reveal the presence of air bubbles.

Pump vanes wear in normal service and can stick in their grooves when the fuel turns resinous after an extended shutdown. Gear-type pumps should require no routine maintenance other than inspection during the course of engine overhaul. Wear tends to localize on the pump cover. When pump failure, as from a sheared drive gear, is suspected, it is

1 Plug
2 Piston spring
3 Piston
4 Feed pump housing
5 Push rod
6 Tappet
7 Roller
8 Check valve seat
9 Check valve spring
10 Check valve
11 Connector bolt
12 Nipple
13 Snap ring
A From the fuel tank
B To the fuel filter

5-36 Piston-type Chrysler-Nissan lift pump.

1 Piston
2 Piston spring
3 Check valve (discharge side)
4 Check valve (intake side)
5 Priming pump
6 Tappet
7 Camshaft

5-37 Operation of the unit shown in the previous illustration. Marine Engine Div., Chrysler Corp.

1	Cover screw
2	Lock washer
3	Pump cover
4	Gasket
5	Inlet valve
6	Outlet valve
7	Valve retainer plate
8	Retainer plate screw
9	Diaphragm and pull rod
10	Diaphragm spring
11	Pump body
12	Rocker arm pin
13	Pin retainers
14	Rocker arm
15	Spacing washers
16	Rocker arm link
17	Rocker arm spring

5-38 AC diaphragm-type lift pump used on Bedford trucks.

sometimes possible to establish that the gears turn by inserting a fine wire into the outlet and cranking the engine. The wire should vibrate from contact with the spinning gears.

Check valves are the weak point of reciprocating (piston or diaphragm) pumps. Most check valves can be removed for cleaning or replacement. When this is not possible, a piece of wire inserted into the fuel entry port should unstick the ball, at least temporarily. Diaphragms should be changed periodically and the housing cleaned every 200 hours or so of operation (Figure 5-39). When installing these pumps, it is good practice to bar the engine over so the pump lever rides on the lower part of the cam.

Fuel filters and water separators

Most engines incorporate three stages of filtration—a screen at the tank and paper-element primary and secondary filters. Filters are rated by the size of particle trapped—a typical installation would employ a 30-micron primary filter and a secondary

1 Cover
2 Pulsator diaphragm
3 Pump body

5-39 Lift pump inspection. Ford Industrial Engine and Turbine Operations

of between 10 and 2 microns. A water separator is often built into the primary filter (Figure 5-40), which also may include a connection for fuel return and an electric heater to prevent fuel gelling and waxing in cold climates.

When a filter element is changed, entrapped air must be bled as described in Chap. 4.

Filter agglomerator head

Out

In

Filter paper element

Transparent bowl

Sedimeter chamber

Drain plug

5-40 Cartridge-type fuel filter with water separator.
GM Bedford Diesel

6
CHAPTER

Electronic engine management systems

Modern diesels respond to their surroundings, engage in two-way communication with their operators, and have a degree of self-diagnostic ability. This chapter discusses how this awareness is accomplished and how it sometimes fails.

WARNING: Exercise extreme caution when working on these systems. Injector voltages can be lethal. Wear rubber-soled boots, safety goggles or face shields, gloves, and protective clothing. Relieve the residual fuel pressure that remains after engine shutdown before loosening fittings. At 100 psi, diesel fuel penetrates the skin; the 30,000-psi pressures used in modern fuel systems strip flesh to the bone. Needless to say, do not test for leaks with your fingers. If skin penetration has occurred, seek immediate medical help.

Tools and resources

Diagnosing electronic control systems is an exercise in data accumulation. The more information you have, the easier the work becomes. While diagnostic systems and the scanners needed to access them have a degree of standardization for passenger cars, heavy truck, marine, agricultural, and industrial engines require dedicated tooling. For example, the $2500 TEXA Universal Diesel Truck Laptop Tool and Scanner retrieves trouble codes, raw sensor data, and enables some component tests for most heavy commercial trucks and many agricultural engines, as well as recent American automobiles and pickups. Dealer-level scanners for similar applications cost upwards of $6000.

Most 1994 and all 1996 American passenger cars have Onboard Diagnostics II (OBD-II). California requires OBD-II for diesel trucks that weigh 10,000 lb and less. Nationally the upper weight limit for OBD-II diagnostics is 6000 lb. Basic OBD-II trouble code readers can be had for as little as $20. Working mechanics and serious DIYers need something like a $250 Auto Enginuity ST06-USB that plugs into a Windows XP / Vista / Windows® 7 / 8.0 / 8.1 / 10 personal computer. Then you need to spend another $200 for what the company calls an "enhanced interface" for the make of vehicle you service. Interfaces are available for Ford, General Motors (GM), Chrysler, and popular imports. Armed with the correct interface, the scanner is no longer a passive observer, but it can exert control over the onboard computer to test individual components and subsystems.

Ford Power Stroke 6.0L and 7.3L pickups use a mix of first-generation OBD and OBD-II and require a compatible scanner, such as an Auto Enginuity ST06 with the suitable interface. The same holds for European diesel passenger cars. The protocol for 2008 and later marine engines is OBD-M. While similar in concept to OBD-II, a dedicated scanner is required.

The tool of last resort is an oscilloscope that converts what scanners report as numbers into visual images (Figure 6-1).

A single-vehicle, one-year subscription to ALLDATA.com costs $26.95 and provides diagnostic trouble codes, diagnostic flow charts, technical service bulletins (TSBs), and repair procedures for the car or truck in question. It can be the best $27 that you spend.

Contributors to Internet diesel forums often upload factory manuals or at least the fuel section portions of the manuals. Used automotive, industrial, and marine shop manuals can sometimes be found on eBay. Manuals for recent engines carry a price tag of $200 or more. However, vehicle manufacturers are legally obligated to

6-1 Serious DIYers and professional mechanics need an oscilloscope such as a 4425 series PC-based PicoScope. The USB-port–powered dual-channel instrument features 16-bit resolution, 5-MHz bandwidth, and a 16-MS buffer memory.

provide short-term access to their repair information. Subscription information can be had at:

BMW	https://www.bmwtechinfo.com/
Chrysler	https://www.techauthority.com/
Ford	https://www.motorcraftservice.com/
Isuzu	http://isuzusource.com/v2/index.php
GM	https://www.motorcraftservice.com/
Volkswagen*	https://erwin.vw.com/erwin/showHome.do

*Erwin also has sites for SEAT, Audi, and Skoda.

The National Highway Safety Administration lists vehicle recalls at https://www-odi .nhtsa.dot.gov/owners/SearchSafetyIssue

Analog and digital

Electronic engine management systems transfer data by means of analog or digital voltages. Analog signals vary over time, as shown in Figure 6-2. The instantaneous value—the voltage at any moment in time—represents data. Voltage spikes, bad grounds, or loose connections seriously compromise the ability of analog circuits to act as reliable messengers. Even so, most sensors are analog devices that communicate changes in temperature, pressure, air flow, and other variables by changing their internal resistances. Supply voltage is constant (normally 5V +/− 0.5V) and output voltage varies as sensor resistance increases or diminishes. The computer converts analog data to digital form.

Critical data, such as the accelerator pedal angle, is transmitted digitally. Because digital signals have only two states—ON corresponding to 1 and OFF corresponding to 0—all the computer needs to do is to discriminate between the two states, or *bits*. Depending upon the system, a string of 8, 16, or 32 bits makes up a *word* that conveys data.

6-2 Analog voltage (A) varies incrementally over time. Digital voltage is binary, present or not, one or zero.

6-3 Pulse-width modulation varies signal ON time to control fuel delivery and collect data.
Courtesy Caterpillar, Inc.

A digital signal may also take the form of pulse-width modulation (PWM). As shown in Figure 6-3, the pulse width of the signal is the percentage of ON time divided by OFF time. Injector power supplies are pulse-width modulated as are the signals from throttle position sensors.

Onboard computer

The computer that oversees engine functions goes by various names—electronic or engine control unit or module (ECU or ECM), powertrain control module (PCM), or microcontroller. Following the example of Cummins, Caterpillar, and Detroit, I'll call the computer an ECM.

The central processing unit (CPU) chip contains millions of transistors that control the engine by integrating sensor data and throttle commands (Figure 6-4). The CPU arrives at its decisions based on arithmetic calculations and look-up data stored in *maps* that plot changes in one parameter against two others. For example, engine load and rpm have the major say on injector timing. The CPU could calculate all values, but it's faster to look up the answer on a map.

Memory

Transient data such as ambient air temperature or crankshaft position can be temporarily stored in random access memory (RAM). Unlike RAM, nonvolatile random access memory (NVRAM) retains stored data when power is shut down. Diagnostic trouble codes (DTCs) are among the data stored in NVRAM. Formulas

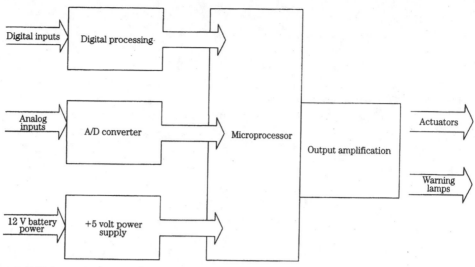

6-4 ECM inputs and outputs.

and diagnostic routines have permanent residence in read-only memory (ROM). Inputs from sensors and outputs to actuators pass through input/output (I/O) ports (Figure 6-5).

Flash memory is a semi-permanent memory guarded by password. Manufacturers use flash memory to update the software, which is never perfect as released. The capacity of the flash memory grows as systems become more complex. For example, the 8-bit Motorola 6800 ECM, used in early Caterpillar control systems and in millions

6-5 The ECM or central processing unit, integrates its memories and data from the external world.

of GM automobiles, had 32 kilobytes (kb) of RAM and 160 kb of flash memory. The 32-bit Adem 2000 currently used by Cat has 256 kb of RAM and 1 Mb (megabyte) of flash memory.

In order to interact with the external world, an ECM requires peripherals, such as a converter to translate analog signals from temperature and pressure sensors to a digital format. Driver circuits, which may be housed separately from the ECM, provide the amperage for injectors, fan motors, turbo wastegates, and other power-hungry actuators. Timers, controlled by oscillating crystals, give the ECM the precise time reference needed to orchestrate injection timing and duration with inputs from the crankshaft position sensor.

EFI systems

Modern electronic fuel systems come in two varieties—common-rail and electronic unit injection.

Common-rail systems

Since the 1980s, engineers realized that an updated common-rail system, using electronically controlled fuel injectors and ultra-high fuel pressures, was the way forward for high-speed diesel engines (Figure 6-6). Two Nippon Denso engineers—Shokei Itoh and Mashiko Miyaki—were responsible for the first commercial common-rail system, which appeared on 1995-model Hino Rising Ranger trucks. Meanwhile Fiat and its subsidiary Magnetic Marelli Power Trains were working on common-rail systems for passenger cars and light trucks. By the late 1990s, a prototype was turned over to Robert Bosch GmbH for production. The first-generation 1350-bar (19,575-psi) Bosch CR made its debut in 1997 on Alfa Romeo and Mercedes-Benz high-speed touring cars.

Subsequent Bosch developments were rapid:

- 1999—1st generation, truck 1480-bar system (Renault).
- 2001—2nd generation, passenger-car 1600-bar system (BMW and Volvo).
- 2002—2nd generation, truck 1600-bar system (MAN).
- 2003—3rd generation, passenger-car 1600-bar system (Audi V6). The use of piezo injectors that cycle open and closed within a ten-thousandths of a second reduced exhaust emissions by 20% and boosted power by 5%.
- 2006—4th generation, passenger-car, 1600/2000-bar system developed, but not yet in production. Features include hydraulically amplified injectors that can generate fuel pressures of as much as 2500 bar.

Common-rail systems have become standard for passenger car and light truck engines, and are rapidly spreading to marine and agricultural applications. Most original equipment manufacturers (OEMs) contract with Bosch, Denso, or Delphi, Siemens VDO for turn-key systems, although more specialized engine makers often design their own systems.

6-6 Common-rail systems are recognized by a high-pressure pump that supplies the injectors through one or, for vee engines, two common rails. Robt. Bosch GmbH.

Common-rail injection offers several advantages over traditional ways of moving fuel into the cylinders. Pre-1980 engines ran with fuel pressures of 6000 psi and sometimes less. Current CR systems develop peak pressures of 2000 bar, or 29,000 psi. High delivery pressures work in concert with multi-holed injector nozzles to atomize the fuel into fog-like particles that penetrate deep into the combustion chamber. Combustion is more efficient and exhaust emissions are reduced.

What makes CR high delivery pressures even more useful is that they can be created at any engine speed. Older unit injector (UI) and pump-line-nozzle (PLN) systems increased pressure linearly with engine rpm. The faster the engine and pump run, the greater is the fuel pressure. Control of fuel pressure provided by CR systems flattens the torque curve and provides another weapon in the struggle against exhaust emissions. In addition, the fuel rail acts as an accumulator—a kind of hydraulic spring—to hold fuel pressure nearly constant as the injectors snap open and close. This relatively constant pressure averages out the torque loads on the injector pump and its drive mechanism. PLN pumps were called "jerk pumps" with reason.

CR high-pressure injection pump

The Bosch CP3 high-pressure injection pump (HPIP), used on 2001-'10 GM and Chevrolet Duramax (LB7, LLY, LBZ, and LMM), 24-valve 5.9L and 6.7L 2003–10 Dodge Ram, Jeep 2.8L, and by thousands of hotrodders, is the closest approximation we have to a universal injector pump. Kits are available to add a second CP3 to vehicles originally equipped with a single CP3 or to the replace the CP4 on 2011 and later Duramax systems.

Material that follows relates to the CP3 as used on the Cummins B-series Chrysler Ram trucks. Chrysler Jeep and GM Duramax CP3 applications differ in detail.

A geared pump under the finned cover on the back of the CP3 boosts the inlet pressure created by the lift pump to approximately 80 psi. Because pressure and flow rates increase with engine rpm, output from the geared pump goes to a cascade valve for regulation.

Cascade valve   The cascade valve (also known as the COV for cascade overpressure valve) is located on the front of the CP3 where it is accessible for removal. The valve distributes fuel for pump lubrication, regulates fuel pressure to the fuel control actuator (FCA), and diverts surplus fuel through a return line to the tank. The internal fuel pressure can be as high as 180 psi for CP3 5.9L applications. The valve consists of a spring-loaded spool that progressively uncovers three spill ports as determined by inlet fuel pressure as follows:

Stage 1   During cranking, the geared pump develops about 40 psi, which is not enough to overcome cascade-valve spring tension. Fuel flows through the center port to spill port to lubricate the pump and to bleed entrapped air back to the tank.

Stage 2   When fuel pressure is greater than about 70 psi and less than 180 psi, the spool moves to open the second spill port for lubrication. This pressure is reached during normal operation.

Stage 3   If gear-pump pressure exceeds 180 psi, the spool spring compresses to open the third spill port that sends the surplus fuel back to the tank. Lubrication ports remain open.

Cascade valves rarely give trouble. If the valve is changed, be careful not to lose the ball. Assembling the valve without the ball will destroy the CP3 through overpressure.

Fuel control actuator   The computer-controlled FCA, also known as the fuel pressure regulator (FPR), fuel metering assembly and by other names, mounts on the rear of the CP3 to regulate how much fuel goes to the high-pressure pump. The FCA remains inactive for as long as 30 seconds during cranking to permit the rail to receive full fuel pressure. Once the engine starts as indicated by the crankshaft position sensor and the fuel-rail pressure sensor, the ECM sends a pulse-width signal to the FCA to keep pump pressure within limits. If the ECM decides that more rail pressure is needed, it energizes the FCA solenoid to increase fuel delivery to the pump. If rail pressure is excessive, the ECM extends the solenoid "off" period.

High-pressure pump section   The CP3 high-pressure section consists of three variable-displacement plungers operated by a three-lobe camshaft. Unlike many earlier injector pumps, the CP3 merely delivers pressure; the injectors determine the timing and amount of fuel delivery. Consequently the pump, which operates at engine rpm, is not timed to crankshaft position. However, gear-driven CP3s should be phased to generate peak pressures during the engine compression stroke. Failure to do so results in gear rattle. While the shop manual describes the procedure, most DIY mechanics phase the pump by trial and error.

Spring-loaded, automatic inlet valves supply fuel to the pump barrels. The valves close as the plungers compress the fuel. Further plunger movement opens the ball-type outlet valve to send fuel to the rail. The CP3 and other modern common-rail pumps generate 4000 to 5000 psi for starting.

HPIP malfunctions

HPIPs eventually fail to produce adequate pressure and/or volume. Normally, the loss of efficiency proceeds slowly over the life of the engine. A scan tool will report output pressures. Output volume can be determined by disconnecting the discharge line at the pump, making up a hose and a catch can to the pump outlet, and cranking the engine for a specified number of seconds. In the absence of supply-side problems, fuel delivery volume should fall within factory specs. HPIP aging rarely affects other fuel system components. Catastrophic failure is another matter.

A single fill up with water, gasoline, diesel particulate fluid or contaminated fuel can disintegrate the pump. Waste vegetable oil can have the same effect. Should this happen, metal fragments permeate the entire fuel system. Cleaning, with the possible exceptions of the fuel rail and tank, is not possible. The pump, injectors, fuel rail, and all plumbing must be replaced often at the owner's expense. Figure 6-7 illustrates the Bosch CP4 pump, which unfairly or not, has a reputation for fragility when using U.S. ultra-low sulfur fuel. An early sign of failure is an accumulation of thick oil and metal fragments on the pressure regulator as shown in Figures 6-7 through 6-10. Other symptoms are a sudden loss of pressure, knocks or other abnormal noises and, most tellingly, yellow "glitter" in the fuel filters.

These problems can, for the most part, be avoided by refueling at busy truck stops and by frequent changes of the fuel filters. Do not wait for the warning light to service the water trap and, if the fuel appears cloudy, immediately drain the system.

Common-rail injectors

Two types of injectors are used on CR systems.

Solenoid-actuated   Figure 6-11 illustrates a multihole, solenoid-operated injector of the kind most often encountered. The spring-loaded needle is poised between the downward force of fuel pressure acting on its flat upper surface and the upward force of fuel pressure on its tapered shoulder. When the injector is at rest, the spring wins out and the needle seats to prevent fuel injection.

When the solenoid is energized, its armature moves upward to open the ball-type spill valve. Fuel flows past the ball valve and into the return line to the tank. This action relieves the pressure against the upper surface of the needle. The pressure balance shifts and pressure on the tapered shoulder of the needle overcomes spring tension. The needle lifts off its seat and injection commences.

Pressure
regulator

6-7 The CP4 pressure regulator is adjacent to the piston assembly and can usually be disassembled for inspection with the pump in place. Robt. Bosch GmbH.

6-8 A clean regulator.

6-9 Glitter-impregnated sludge on the regulator means catastrophic pump failure.

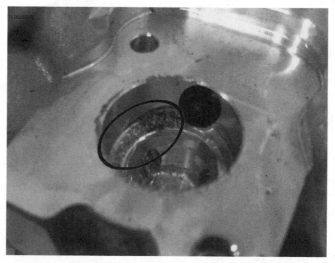

6-10 As opposed to normal wear, pump fragmentation contaminates the entire fuel system.

6-11 Solenoid-operated injector. Courtesy Bosch.

Solenoids require large amounts of power. In order to overcome armature inertia and the drag of heavy fuel oil, Bosch solenoid injectors receive a 50V, 20A jolt of power during the opening phase. Once the armature is retracted, battery voltage holds it in place.

WARNING: injector voltage can be lethal.

Solenoid injector diagnosis   Symptoms of injector malfunctioning are:

- Slow or no starting.
- Rough idle that may be accompanied by knocking.
- White or bluish white smoke in the exhaust results from cold combustion. If the smoke persists for a minute or two after starting, the injectors could be involved. Note that the diesel-particulate filters on late-model engines must be disconnected for exhaust smoke to be visible.
- Reduced engine performance.
- Increased fuel consumption.
- Failure to pass an emissions test.
- Injectors that leak fuel into the cylinders dilute the oil, as evidenced by smell and a rising oil level on the dipstick. The leaked fuel burns pistons and, in severe cases, results in hydrolock.

WARNING: CR fuel systems retain pressure after the engine is shut down. How this pressure is relieved in preparation for opening high-pressure fuel lines varies with make and model. *But the pressure must be relieved for your own protection.* Do not forget that injector voltage can be lethal.

Using a scan tool, disable one injector at a time. Diminished knock, vibration, or exhaust smoke suggest that the associated injector is bad. However, this test is not definitive; fuel can enter the cylinder if the injector ball valve leaks. To be certain, block off the injector supply line with the appropriate tool. For example, the Cummins Dodge 2003-07 5.7L needs a Tamer X or Miller 9011 block-off tool, designed to withstand rail pressure.

A professional-level scan tool will also report cylinder balance which is a comparison of the pulse widths for each injector to equalize power from each cylinder.

Check for excessive fuel return. Remove and plug the return rails or hoses (Figure 6-12). Replace the return hoses on each injector with a length of windshield wiper hose. Use identical containers for catch cans and idle the engine. If the engine refuses to start, force the issue with a quick shot of brake cleaner in the air intake. After 5 minutes or so, shut off the engine and compare the fluid levels in the catch cans. Injectors that bypass more than the average amount of fuel are bad. No fuel in the return line means that the injector is leaking in fuel into its cylinder.

The definitive way to evaluate injectors is to send them out for professional testing and ultrasonic cleaning. The technician will visually check the spray pattern, listen for the "chirp," verify that the nozzle opening pressure (NOP) is within specification, gauge the precise volume of fuel delivered and that the injector closes without dribble.

Piezo electric Piezo injectors evolved from the flow-control technology developed by Siemens VDO for ink-jet printers. Piezo crystals expand when subjected to voltage and generate voltage when compressed. The latter characteristic is used in cigarette and barbecue lighters.

The growth in crystal volume is minuscule—an 80-μm crystal expands 0.00002 in. when subjected to 140V. Siemens, Delphi, Bosch and other manufacturers stack the wafers to produce a stroke of about 40 μm which is then hydraulically amplified to unseat the spill valve. The wafer pack responds within 80 milliseconds of excitation, or three times faster than the best solenoids, and with an actuation delay on the order of three magnitudes faster. Injection can be divided into as many as five precisely metered shots, each accurate within less than a milligram of fuel. This translates as 3% more power and as much as 15% more fuel economy than afforded by solenoid injectors. Because combustion is more complete, exhaust emissions are significantly reduced.

Piezo injectors operate on the same 1600–2000-bar common-rail pressures used for solenoid injectors. Depending upon the application, injector tips have as many as 10 orifices with diameters of less than 0.1 mm.

Figure 6-13 shows a piezo injector in cutaway view. Wafer expansion opens a spill valve between the high-pressure fuel circuit and the low-pressure return. When the valve opens, pressure above the needle diminishes, while pressure in the nozzle cavity remains nearly constant. The differential unseats the needle to initiate injection.

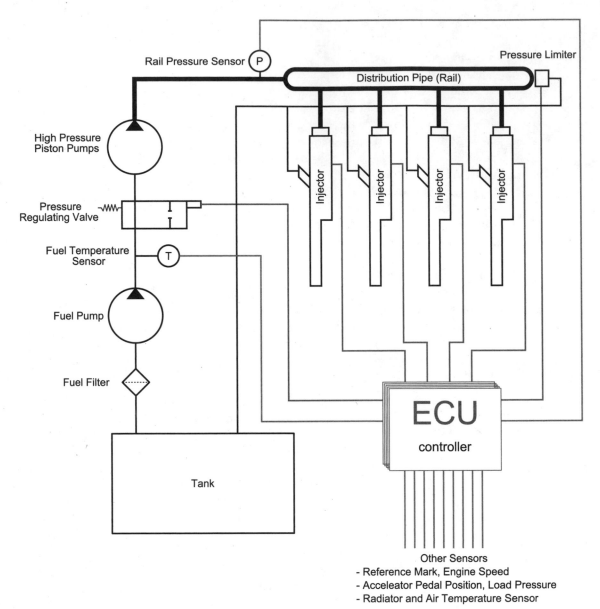

6-12 CR systems orchestrate fuel delivery with an ECU (electronic or engine control unit) that integrates sensor data to determine injector timing, volume, and duration. Note that high pressures exist downstream the high-pressure, pump. A low-pressure line supplies the pump and spill lines from fuel-pressure regulators and injectors recycle fuel back to the tank. Wikimedia Commons, created by Smakkeloep.

6-13 A piezo injector in cross-section, showing the fuel return (1), fuel inlet (2), wafer stack (3) hydraulic coupler (4), spill valve (5), nozzle assembly (6), and injection orifices (7). Courtesy Bosch.

Because wafer expansion is minuscule, a hydraulic coupler extends the stroke. The coupler also functions like a hydraulic valve lifter to compensate for lash created by thermal expansion.

Piezo diagnosis   Piezo injector malfunctions produce the same no start, rough idle, power loss, and/or exhaust smoke systems as conventional injector failure. Definitive testing is the province of professionals who have access to factory-approved tooling and training.

Pressure-relief valve

Because the injection pump delivers more fuel than displaced when an injector fires, fuel pressure in the rail remains nearly constant. The ECU receives pressure reports multiple times a second from sensor on the rail and adjusts pump output accordingly. But the response is not instantaneous, and fuel pressures can reach

60,000 psi during sudden decelerations. A spring-loaded pressure-relief valve then opens to bleed off the excess rail pressure. Relief valves weaken with age and can open under normal rail pressure.

Electronic unit injection

Designers of heavy-duty engines have been reluctant to abandon unit injection, which transfers the responsibility for high fuel pressures from the pump to the cam-operated injectors. Should an injector fail, the engine will continue to run. Confining high fuel pressures to the injectors also reduces the potential for leaks and eliminates the danger of widespread metal contamination that occur if a remote injector pump disintegrates. It is also true that common-rail systems develop pressure pulses when supplying the fuel needs for large engines. These pulses interfere with fuel calibration.

Figure 6-14 illustrates a Bosch EUI used on Volkswagen, SEAT, Skoda, and other European passenger cars. The roller-tipped rocker arm acts on the injector plunger to generate as much as 30,000-psi fuel pressure. Injection begins when the ECU energizes the solenoid to close the spill valve at any point during plunger downward movement. Fuel pressure then lifts the needle to commence injection. A check valve contains injection pressure within the injector body. When the needle and associated plunger retracts, a check valve opens to recharge the injector with 50-psi fuel.

EUI diagnosis   Electronic unit injector (EUI) failure can result in exhaust smoke, misfiring, vibration, and slow or no starting. Identify failed EUIs by disabling one injector at a time with a scan tool or by using the same tool to make a cylinder balance test.

6-14 Bosch unit injector. Engine camshaft (1), plunger (2), solenoid spill valve (3), and nozzle (4). The overhead camshaft eliminates flexible pushrods for precise injector articulation. Courtesy Bosch.

Frequently encountered EUI failure modes are:

- Excessive clearance between the actuating cam and the plunger—adjust injector height to specification.
- Plunger stuck in the retracted position—replace injector.
- Cracked injector barrel or tip—replace injector.
- Fuel leaks—if external, check for cracks in the injector barrel and replace the O-rings. Leaks on the engine end of the injector can burn pistons and, if not corrected, flood the crankcase with diesel fuel.
- Shorted or open solenoid—replace injector.
- Damaged connector—replace pigtail.
- Shorted wiring harness—most wiring malfunctions occur at head-gasket connections—replace head gasket with the OEM part.

Hydraulic/Electronic Unit Injection

HEUI (pronounced "Huey") injectors are the product of a joint venture between Caterpillar and International Truck. Introduced in the mid-1990s, HEUIs were used on Caterpillar, International T444E, DT466E, and 1530E, certain Perkins models, and, most famously the Ford Power Stroke (Figure 6-15). Unlike conventional camshaft-driven unit injectors, HEUIs are actuated by high-pressure crankcase oil. Since injection is no longer tied to camshaft motion, fuel can be injected at any crankshaft angle. At higher-than-idle speeds, injector pressure is independent of engine rpm and can attain values of 30,000 psi.

A HEUI consists of five major components—a solenoid, poppet valve, mushroom-shaped intensifier piston, barrel, and a seven-hole nozzle assembly (Figure 6-16). Four amps energizes the solenoid and 1.5A holds it open.

The solenoid unseats the poppet valve to admit high-pressure lube oil to the upper end of the intensifier piston (6 in Figure 6-16). Oil pressure forces the piston

6-15 HEUI injection system.

6-16 A HEUI in cross-section. Solenoid (1), armature (2), upper poppet seat (3), poppet valve (4), lower poppet seat (5), intensifier piston (6), plunger (7), plunger cavity (8), barrel (9), and nozzle assembly (10). Force multiplication occurs between the oil and fuel ends of the intensifier piston. Courtesy Caterpillar, Inc.

and attached plunger down against fuel trapped in the injector cavity by a ball-check valve. The surface area of the piston is seven times greater than the surface area of the plunger. This size differential multiplies piston force by seven: the 3600-psi oil pressure developed by the 6.0L Ford Power Stroke creates 25,200 psi of fuel pressure. The pressurized fuel unseats a second check valve to enter the nozzle fuel cavity, where it lifts the needle to initiate injection. NOP (nozzle opening pressure) varies with the application. How much fuel the injector delivers depends upon oil pressure and pulse width.

In 1994 Ford introduced the split-shot PRIME (pre-injection metering) HEUI on F-series trucks sold in California. Subsequently PRIME injectors became standard across the product line. These injectors work like other HEUIs, except that the plunger incorporates a radial groove that receives fuel through six bleed ports drilled in the plunger face. As the piston moves downward, the groove aligns with a spill port in the barrel to shunt pressure. The output-side check ball seats and injection ceases. Further piston movement masks the port and injection resumes. The short burst of early injection, known as pilot injection, quells vibration and helps to control oxides of nitrogen emissions.

HEUI diagnosis   HEUI failure is not always easy to diagnose, since rough idle, long cranking periods, and misfires at certain engine speeds can have multiple causes, several of which are associated with the high-pressure lube oil system. It is helpful to check individual exhaust runner temperature with an infrared thermometer. All manifold runners should be very close to the same temperature: a hotter-than-average runner means that the associated cylinder receives too much fuel; a cold runner means insufficient fuel. Either of these conditions point to injector failure.

Poppet-valve wear first appears as rough running on start up. As the oil heats, it thins and injector action returns to normal. Tolerances are extremely tight; the wrong oil, air in the oil supply, or contaminated fuel quickly destroy the plunger/barrel fit to divert fuel back to the tank. Monitoring the return flow while disabling one injector at a time identifies those with excessive returns.

The three external O-rings are critical. The upper O-ring, which is now supplied with steel and elastomer backup rings, prevents lube oil from leaking out of the injector body. The middle O-ring and its supporting rings isolates lube oil from the fuel. Sudden increases in oil consumption can usually be attributed to failure of this middle ring assembly. The lower O-ring keeps fuel from puddling around the base of the injector.

Exert care when servicing injectors and other parts that depend upon O-rings for sealing. Plumbing, and especially fuel-return lines, must be secure. Replace seals whenever connections are disturbed.

Sensors

Because sensors work by physical contact, they can be identified by their location. Fuel sensors must be wetted with fuel, turbo speed sensors mount on the turbocharger, EGR sensors are exposed to exhaust gases, and so on.

Most sensors are analog devices connected to the ECU by three wires. The ground wire completes the circuit to power up the sensor. The second wire carries the 4.5V–5.5V reference voltage (V_{ref}) and third wire the signal, or output voltage (V_o).

The ECU flags a trouble code in the event of:

- Low, high, or no reference voltage.
- Low, high, or no signal voltage.

High reference voltages are hardly ever encountered and indicate that the voltage-regulation circuitry has failed. Low or no reference voltage usually can be traced to a connector with corroded or bent pins. Wiggling the connector sometimes reveals the fault. Abraded or pinched wires short reference voltage to ground.

Low or no signal voltage at the sensor connector means that the sensor has failed. No or erratic sensor voltage at the ECM connector indicates a problem with the wiring harness. Use the wiring color code to keep track of pin connections.

What follows is a discussion of the sensors that most frequently give problems:

Temperature and pressure sensors   Most temperature sensors operate on a 5V supply voltage and have a negative temperature coefficient (NTC), which means that the internal resistance of the sensing element decreases with heat. The higher the temperature, the greater the output voltage. Peak output voltage will be something on order of 4V. Positive temperature coefficient (PTC) sensors work in the opposite way to decrease output voltage as temperature increases for a peak output of 1V or less. Pressure sensors respond like NTC sensors: the higher the pressure, the greater the output voltage value. At maximum pressure the output will be a bit less than the 5V reference voltage.

MAP (manifold air pressure) sensor   An ordinary pressure gauge measures pressures in reference to ambient air pressure. Zero on the gauge is approximately 14.7 psi. But ambient air pressure changes with altitude and temperature. A MAP sensor measures intake manifold pressure against the zero pressure exerted by a vacuum. Consequently, its pressure readings are absolute and reliable.

A MAP sensor may mount directly on the intake manifold or be connected to it with a flexible hose. The sensor has two chambers, one under a nearly pure vacuum (with pressure as low as 0.0005 psi) and the other open to the intake manifold. A diaphragm separates the two chambers. A semiconductor responds to diaphragm movement by altering its electrical resistance. As manifold pressure increases, the diaphragm elongates toward the vacuum chamber. This action stresses the semiconducter element to divert more of the 5V reference voltage to signal voltage.

The ECM used with MAP sensors operates in what is called the Speed/Density mode. That is, the computer integrates manifold absolute pressure, air temperature, and engine volumetric efficiency with engine rpm to calculate oxygen consumption and adjusts fuel delivery accordingly. Bar 1 MAP sensors work at one atmosphere maximum, bar 2 and bar 3 sensors are for turbocharged engines that generate two or three times ambient pressures. Because of the greater span of forced-induction MAP sensors, some precision is lost. Harness connections are critical because at near atmospheric pressures signal voltage is only small percentage of the 5Vref voltage.

MAP sensor failure results in sooty exhaust smoke from overly rich air/fuel mixtures, rough running, and other ambiguous symptoms. The ECM may respond by putting the engine in the limp-home mode. Note that vacuum leaks in the connecting plumbing can mimic MAP failure. These sensors can be expensive and should be bench tested before purchasing another one. To make the test you will need sensor response data (available from the factory manual), a hand vacuum pump and a 5V power supply. Verify that signal voltage rises as specified throughout entire pressure range. Voltage should increase smoothly without flat spots or stutter.

6-17 A Bosch membrane, or hot film, MAF sensor.

MAF (mass air flow) sensor 5V or 12V MAF sensors, found on many light- and medium-duty engines, mount in the intake manifold and measure air flow by its cooling effects on a heated wire or membrane (Figure 6-17). The amperage needed to keep the sensing element at a constant temperature has a direct relationship to air flow and fuel demand.

A malfunctioning MAF sensor trips multiple trouble codes and can result in exhaust smoke or a lag in acceleration. Power, whether from too much fuel or too little, will be reduced. Hot-wire sensors may respond to cleaning with an aerosol intended for the purpose. (Brake and throttle-body cleaners leave a harmful residue.) Cleaning membrane sensors does not seem to make much difference.

If you do not have the specifications for the MAF, an idea of its condition can be had by monitoring the output signal on a warm engine. Voltage or pulse-width should increase steadily with engine speed. Hesitation or momentary loss of signal strength are grounds from rejecting the sensor. Note that malfunctioning exhaust-gas recirculation hardware or intake tract leaks mimic sensor failure.

CPS (crankshaft position sensor) A CPS reports engine No. 1 piston position by means of a marker tooth on the crankshaft gear or sprocket. One or more gear teeth will be missing or undercut in a way that registers on the sensor coil as a change in magnetic properties (Figure 6-18). Inductive sensors usually have a two-pin connector; the newer Hall effect sensors have power supply, ground, and signal pins.

CPS failure results in engine shutdown, rough running, or recourse to the limp-home mode. As symptoms worsen, the engine can refuse to start or start only when cold. Some Caterpillar and Mercedes applications respond to CPS failure by substituting data from the camshaft position sensor.

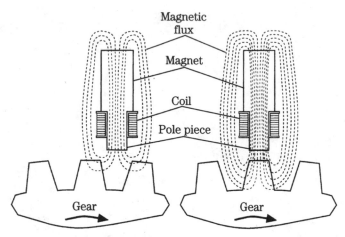

6-18 CPS sensors use a missing, modified or a magnetically charged crankshaft gear tooth to report the position of No. 1 piston. Dynalco Controls, a unit of Main Controls Corp.

Camshaft position sensor (CMP)   CMP reports camshaft speed and position. These inductive or Hall-effect sensors function like crankshaft position sensors and have similar failure symptoms. Figure 6-19 illustrates how an oscilloscope is connected to test square-wave sensors. Without such a tool, test by substitution of known good parts.

Accelerator pedal position (APP)   Because of safety concerns, APPs have redundant sensing elements. Old-style APPs used as many as three potentiometers, any of which would enable the engine to accelerate past idle. Newer types employ multiple magnetic sensors.

Most APPs are analog devices that report accelerator position by changes in output voltage. Typical readings are 10% of the 5V reference voltage at idle and 85% at wide-open throttle. Pulse-width APPs exhibit similar percentages of full pulse width.

Failure tends to be progressive with flat spots during acceleration that become more frequent and end as "dead pedal."

Actuators

The primary actuators are the electronic injectors. Another almost universal actuator is the electronic governor that can be all-speed for vehicles, min-max for generators, and full-power for other applications. The latter often has a timer associated with it to prevent engine damage.

Other actuators control fuel pressure, turbo boost, and exhaust back pressure.

6-19 An oscilloscope showing wave forms for both the crank and camshaft sensors can detect multiple faults including loss of synchronization. The camshaft sensor signal should occur every 360° of crankshaft rotation. If not, the engine is mistimed or the sensor and possibly its associated wiring is at fault. Pico Technology.

Lift pump

A lift pump (also called a supply or transfer pump) supplies fuel at low pressure to the injector pump. Older engines often used a diaphragm lift pump driven off the engine or injector pump camshaft. Most recent engines have an electric vane- or gear-type pump in the tank or as close as practical to it.

GM did not include a lift pump on Duramax 2001-10 engines, which means that the CP3 injector pump must feed itself. As one might expect, vacuum-induced air leaks in the supply line are common. There is also reason to believe that injector pump longevity is adversely affected by the lack of a lift pump. Certainly this is true if air leaks are present.

Original equipment lift pumps are not without problems, as 2003–04 Dodge Ram owners can attest. These engines have their lift pump mounted on the driver's side of the engine block, remote from the tank. Air leaks cause the fuel-cooled VP44 injector pump to overheat and fail. Chrysler supplies a kit to replace the lift pump with the in-tank unit specified for late-model trucks. However, many owners prefer to make the update with an aftermarket pump.

The least-expensive pumps have brushes that eventually wear out. The shaft seal is also a sacrificial item. Another issue is loss of pump pressure as the fuel level drops below a quarter full. Light truck tanks tend to be narrow because of the limited space between the outboard frame rails and the driveshaft. During acceleration and braking, fuel collects at the ends of the tank to uncover the pickup tube, which then draws air. The quarter-tank pressure problem can be eliminated by welding a deeply dished sump in the bottom of the tank. Beans Diesel Performance is a popular source of these sumps.

For peace of mind, an aftermarket lift pump should be capable of flow-through. That is, pump failure should not deny fuel to the injector pump. Nor should the replacement pump have more output than the engine needs. High-capacity pumps merely recycle surplus fuel back to the tank, heating it in the process.

FASS, AirDog, BD Diesel, and several other manufacturers make quality lift pumps with filters, adjustable pressure valves, automatic air bleeds, and other features. Prices average around $700, although installation and options add to the cost.

Figures 6-20 through 6-22 illustrate installation of an Aeromotive pump for 2001–10 GM Duromax engines. The 130 gph @ 10 psi unit has a seal-less and brushless motor, and a 2-μ Caterpillar filter. Pressure is adjustable.

Aeromotive suggests that a professional install the pump. If you opt to do the work yourself:

- Run the fuel tank nearly dry.
- Disconnect the negative battery cables and then the positive cables making sure that the cables cannot spring back into contact with the battery terminals.
- Support the vehicle on a hard surface with solid support provided by jack stands. Do not work under a hydraulic jack.
- Extinguish any sources of ignition and work in a well-ventilated area. Have an approved fire extinguisher nearby.

6-20 Aeromotive PN 11801 pump connection points.

6-21 The cutout for the fuel sender provides access to feed and return tube hardware. In addition, the fuel-gauge arm must be free to move and not bind against the sides of the tank.

Three requirements determine where the holes for the pickup tube and return tube will be drilled. The first requirement is that the holes must be in a place where there is sufficient vertical space between the tank and underside of the bed for the tubes. Make this determination before dropping the tank.

The second requirement is that the movement of the fuel-gauge arm must not be compromised (Figure 6-21). Locate the holes so that the nuts that go on the underside of the tubes are close enough to the fuel gauge-sender cutout to be reached with a backup wrench.

6-22 Aeromotive pump installed on a chassis rail.

Drilling holes in the tank pose the risk of fire and explosion. Some people drill diesel tanks without problems. Others blow themselves up. All that can be done is to minimize the risk:

- Drain the tank and wash or steam out all traces of fuel.
- Flood the tank with CO_2 from dry ice or a large fire extinguisher.
- Coat the area to be drilled with grease to catch the swarf.
- Drill the holes with a sharp bit and a cordless (non-sparking) drill motor.

What remains is straightforward—the pump is mounted on a chassis rail with a bracket included in the kit, hose connections are made up and a wiring connected.

The brief descriptions of engine management systems that follow do not substitute for factory documentation, but should give readers a sense of the terrain.

EMS troubleshooting

Table 6-1 lists the most common malfunctions that diesels are heir to. But engine management systems differ and must be approached on their own terms. Factory shop manuals, service bulletins and parts catalogs are essential.

Table 6-1. Basic troubleshooting—EMS-equipped engines

Item to check	Action
Trouble codes	Scan active and past trouble codes, and make the necessary repairs.
Fuel contamination	Look for evidence of water, macrobiotic growth, and solids.
Fuel specific gravity	Test with a hydrometer. No. 2 diesel should have a specific gravity (S.G.) of 0.840 or higher. Use of No. 1 diesel can cut power by as much as 7%. Mixing Nos. 1 and 2, a common practice in cold climates, costs a proportional amount of power.
Fuel shutoff valve— mechanical or solenoid	Make sure the valve opens fully.
Fuel temperature	Although fuel temperature varies with the application, 85°F is a reasonable figure. Every 10° increase in temperature above 100°F cuts power by 1%. Restrictions—crimped lines, clogged filters, sharp bends—increase pumping loads and heat. Verify that the fuel heater functions normally.
Restrictions in fuel filters and associated components	Change fuel filters. Look for restrictions at points where fuel flow is disturbed with particular attention to the water separator, check valves, and flow meter. Examine the lines for physical damage.
Tank vent restriction	A clogged vent causes the tank to develop vacuum as fuel is consumed.
Fuel leaks	Check for fuel leaks, which occur most often on the high-pressure side of the system.

(Continued)

Table 6-1. Basic troubleshooting—EMS-equipped engines (*Continued*)

Item to check	Action
Air intrusion	Air usually enters on the suction side of the transfer pump. Look for bubbles in the injector return line and in the filter-to-injector-pump line.
Turbo boost	A rubbing turbo wheel, bent blades, oil leaks, or failure of the VGT to function properly reduce turbo boost. Potential leak sites include the exhaust plumbing to the turbo and, on the outlet side, the intake manifold-gasket and aftercooler. Replace the air filter unless obviously new. Excessive turbo boost with normal fuel delivery indicates retarded injector timing, which works against the additional boost to reduce power, increase fuel consumption, and raise exhaust temperatures. These high temperatures accelerate exhaust-valve and turbocharger failure.
Air charge temperature	Consult the shop manual to determine the permissible charge-air temperature. In normal operation, an efficient aftercooler should hold the temperature rise to no more than 30°F over ambient. Maximum temperatures, as generated during full-throttle dyno tests of turbocharged engines, should be about 325°F without an aftercooler, 245°F with a jacket-water aftercooler, and 150°F with an ambient air cooler.
Injector-pump timing	Check injector-pump timing against manufacturer's specs with a dial indicator as described in Chap. 4.
Misfiring cylinders	Locate weak or misfiring cylinders by shutting down one injector at a time with the appropriate factory tool while the engine is running. Replace the associated injector(s). If that does not restore cylinder function, make a compression test as described in Chap. 4.
EGR function	Refer to manufacturer's manual for test procedures. Remove carbon deposits from valve and manifold inlet porting. Check turbo-pressure control and monitoring devices.
Exhaust restriction	Check exhaust piping for damage; verify back pressure against manufacturer's specs. Exhaust back pressure readings of more than 27 in./H_2O for turbocharged engines or 34 in./H_2O for naturally aspirated engines are cause for concern.
Parasitic loads	Check for high current draws, air leaks in compressed-systems, and other parasitic loads that reduce engine power output.
Governor seals	Look for damage to factory-sealed adjustments that, for example, can indicate that governed idle speed has been increased in an attempt to boost full-throttle power output. Using an accurate tachometer (not the control-panel unit), compare idle speed against the factory specification.

Caterpillar EMS

Figure 6-23 illustrates the EMS used on Caterpillar 3112, 3176, 3406E, and 3500 engines. It is a relatively simple system without the bells and whistles mandated by current emissions regulations. The ECM mounts on the engine block, which reduced the electromagnetic radiation given off by the harness, simplifies packaging and enables the computer to be cooled by fuel. A major engineering effort was required to isolate the electronics from heat, vibration, and cleaning solutions.

All sensors, with the exception of the oil pressure sensor, input data for fuel allocation. Abnormally low or high sensor readings cause the ECM to set one or more trouble codes, which are retrieved by connecting a scan tool to the J1922 data link connector.

6-23 Cat 3406E engine management system in block diagram.

The only computer-controlled actuators in this system are the EUIs, cooling fan, and cruise control.

A more sophisticated system used on C-10, C-12, and C-15 truck engines is illustrated in Figure 6-24. Sensor and actuator functions for this and earlier Cat systems are

6-24 C10, C12, and C15 engine management system. Courtesy Caterpillar, Inc.

Table 6-2. Component overview—caterpillar EMS

Nomenclature	Description
Electronic unit injector (EUI)	Unit injector with injection timing and volume controlled by a solenoid slaved to the ECM.
Atmospheric pressure sensor	Absolute pressure sensor measuring pressure from 0–16.8 psi. Supplied with 5V.
Oil pressure sensor	Absolute pressure sensor measuring oil pressure from 0–165 psi. Data communicated to service tools is absolute pressure less atmospheric pressure.
Coolant temperature sensor	Thermistor not requiring an external voltage supply. Used to monitor engine operation during cold starts and to reduce power output when coolant temperature is excessive.
Inlet manifold air temperature sensor	Thermistor not requiring an external voltage source. Provides data for control of the inlet air heater, cold idle, and cooling fan.
Fuel temperature sensor	Used to adjust fuel delivery and to limit engine power when fuel temperatures exceed 86°F (30°C). Maximum power reduction occurs at 178°F (70°C). Higher fuel temperatures flag a trouble code.
Crankshaft and camshaft position sensors	Report engine rpm and timing data for fuel injection. Normally, the ECM monitors both the camshaft and crankshaft position sensors during cranking and the crankshaft sensor while running. If either sensor fails, the engine will start and run using data supplied by the other.
Turbo wastegate actuator	A solenoid that regulates turbo boost.

listed in Table 6-2. Figure 6-25 shows the location of EMS components on C-10 and C-12 engines. C-15 sensor location is similar, except that boost and manifold-pressure sensors are on the left side forward, rather than on the right.

Caterpillar 3126 HEUI

Figures 6-26 and 6-27 illustrate the HEUI injection system as adapted to the 3126B truck engine. A seven-piston swash-plate pump, mounted forward on the driver's side of the block, supplies oil pressure to the injectors through a cylinder-head passage. Oil returns to the sump through ports under the valve cover. Rebuilt oil pumps, also used by Ford on the 7.3L, International and Perkins, are available on the aftermarket. Injector operation has been described earlier under the "Hydraulic/Electronic Unit Injection" heading.

The injector actuation pressure control valve (IAPCV) regulates oil pressure by mean of a solenoid-driven relief valve (Figure 6-28) that takes it cues from a pressure sensor in the cylinder-head gallery. Actuating oil pressures range from 870 psi during cranking to as much as 3500 psi under full load.

Left Side View

A

Front Right Side View

B

Front Top View

6-25 Cat EMS component location on the left (A) and right (B) sides of C10 and C12 Caterpillar engines.

(1) HYDRAULIC PUMP
(2) OIL FLOW TO ENGINE
(3) OIL FILTER
(4) ENGINE OIL PUMP
(5) OIL COOLER
(6) IAP SENSOR
(7) INJECTORS
(8) FUEL SUPPLY RAIL
(9) FUEL PRESSURE REGULATOR
(10) IAP CONTROL VALVE
(11) FUEL FILTER
(12) FUEL TANK
(13) BACK OF CAM GEAR
(14) SPEED TIMING SENSORS
(15) ECM
(16) BOOST PRESSURE SENSOR
(17) ACCELERATOR PEDAL
(18) ACCELERATOR PEDAL POSITION SENSOR
(19) BATTERIES
(20) COOLANT TEMPERATURE SENSOR
(21) DATA LINK
(22) EXHAUST BRAKE RELAY
(23) INLET AIR TEMPERATURE SENSOR
(24) TRANSMISSION RELAY
(25) VEHICLE SPEED SENSOR
(26) INLET AIR HEATER RELAY
(27) INLET AIR HEATER LAMP
(28) FAST IDLE LAMP
(29) ENGINE CHECK LAMP
(30) SPEEDOMETER & TACHOMETER
(31) CRUISE ON/OFF AND SET/RESUME SWITCHES
(32) PTO ON/OFF AND SET/RESUME SWITCHES
(33) SERVICE BRAKE SWITCHES
(34) NEUTRAL & CLUTCH SWITCHES
(35) FUEL TRANSFER PUMP

6-26 The Cat 3126 HEUI was the pattern for the 7.3L Ford Power Stroke and International Truck fueling systems.

126

6-27 Cat 3126 fuel plumbing incorporates galleries in the cylinder head for fuel delivery and return, tubing between major components and return hoses. Fuel filter supply tube (1), lift pump to filter (2), pressure regulator (3), return line to fuel tank (4), lift pump connection (5), filter (6), and fuel supply tube (7).

6-28 The IAPCV regulates the pressure of the injector actuating oil. Parts labeled are the discharge port to injectors (1), electrical connection (2), valve seal (3), poppet valve (4), and valve return spring (5). When energized as shown here, the solenoid extends its armature to move the poppet valve to the left. This action causes the poppet valve to block the drain port (located just above it in the drawing). The actuating oil now comes under full pressure. When the ECM cuts power to the solenoid, the poppet valve moves to the right in response to line pressure. This action opens the drain port to bleed off pressure. Courtesy Caterpillar, Inc.

Should the IAPCV fail, replacing the O-rings can sometimes save the cost of a new part. Caterpillar supplies a special tool—a kind of flare-nut crows-foot wrench—for this purpose. Torque to the IAPCV to 37 lb/ft.

Caterpillar EUI

Both the 3100-3500 and C series engines employ EUIs that combine mechanical actuation with electronic control over fuel volume.

Roller-tipped rocker arms actuate the 28,000-psi injectors (Figure 6-29). The solenoid valve on the left of the drawing opens the spill/fill port to admit fuel to the injector barrel during the period of plunger retraction. A spring-loaded check valve, located near the injector tip and set to open at 5000 psi, remains closed during the fill process. Injection can be initiated at any time after the plunger starts its downward travel. But until the ECM signals the solenoid valve to close fill/spill port, fuel merely cycles through the injector as a coolant and as a purge to remove trapped air. Upon signal, the solenoid valve closes the spill ports, trapping fuel in the injector barrel. Further downward movement of the plunger raises the fuel pressure sufficiently to overcome spring tension acting on the check valve. The check valve then opens and injection begins.

Although the ECM has final control over injector timing, the onset of pressure rise depends upon the clearance between the rocker arm and the plunger. As shown in Figure 6-30, dealer mechanics make this critical adjustment with a PN J 35637 height gauge. You can also use the height gauge feature on digital calipers.

Fuel pressure

The piston-type transfer pump runs off an eccentric on the back of the injector oil pump. Nominal injector pressure is 65 psi at the pump. When the measurement

6-29 Caterpillar EUI in cross-section.

Electronic
unit injector
(EUI)

Height gauge
J 35637

Gauge hole

6-30 A Caterpillar-supplied gauge is used to set injector height relative to the camshaft.

is made downstream of a new filter, expect to see a pressure drop of about 5 psi. If pressure is low, clean the inlet screen on the pump and replace the fuel-pressure regulator with a known good unit. If the problem persists, replace the transfer pump.

Boost pressure

The amount of turbo boost relative to engine speed and load depends upon the application, with the specification given for 28.8 in./Hg dry barometric pressure, 77°F (25°C), and 35 API gravity fuel. If the barometric pressure is higher and/or the temperature lower, boost pressure will be higher than specified. By the same token, heavier fuel increases turbo output. The wastegate is nonadjustable.

Low manifold pressure combined with high exhaust temperature suggests that the air filter is clogged or that the aftercooler, pressure-side turbo plumbing, or intake manifold leaks.

Throttle position sensor (TPS)

The TPS is located under the accelerator pedal and generates a pulse-width signal that can be read with a Fluke 88-5 multimeter or the equivalent and a breakout box (Figure 6-31). The breakout consists of three 12-in. long No. 10 AWG wires that bridge the harness connectors and three 6-in. long No. 12 AWG leads that make up to the meter. Follow this procedure:

1. Using Caterpillar PN 1U5804, crimp Deutch DT 04-3P0-E008 male connectors to one end of each of the three long wires and DT 04-3P-E003 female connectors to the other ends.
2. Make up DT 04-3P-E003 female connectors to one end of each of the three short wires. These connectors plug into the meter.
3. Solder the three short wires to the centers of the longer wires.

6-31 A breakout box must be compatible with the vehicle harness connector. Courtesy Caterpillar, Inc.

Set the meter to measure pulse-width percentage and, with the ignition switch ON depress the accelerator pedal. Pulse width varies with model, but should be in the neighborhood of 80% or 90% at full throttle and 10% or 20% with the pedal released. Keeping a spare TPS on hand is good insurance.

Diagnostic Trouble Codes

Table 6-3 lists the most critical DTCs. The parameter identifier (PID) is a two- or three-digit Society of Automotive Engineers (SAE) numerical code assigned to each component. FMI is a failure mode identifier that describes the kind of failure detected, and Flash lists the number of blinks the code triggers on the check engine light. Some DTCs are merely informational and do not reflect problems with engine operation. Active codes, that is, those that turn on the diagnostic lamp and keep it on, mean that the malfunction requires immediate attention. The lamp should go out when the malfunction is corrected. Intermittent malfunctions, often caused by loose harness connectors or bad grounds, cause the lamp to blink and go out. These codes are logged in memory where they can be removed with an appropriate scanner. Intermittent codes that are logged repeatedly need investigation.

Ford 7.3L Power Stroke

Ford sold more than two million 7.3L Power Strokes during a production run that lasted from the 1994 to the 2003.5 model years. People took out second mortgages to buy these trucks. While fuel economy could have been better, the under-stressed engine seems to run forever. It has a B10 200,000 mile reliability rating, which means

Table 6-3. A partial list of Caterpillar HEUI trouble codes

PID and FMI	Flash code	Description
01–11, 2–11	72	Cylinder 1 or 2 fault
3–11, 4–11	73	Cylinder 3 or 4 fault
5–11, 6–11	74	Cylinder 5 or 6 fault
22–13	42	Check timing sensor calibration
41–02, 41–03	21	8V supply above or below normal
42–11	18	IACPV fault
64–02, 64–11	34	Loss of engine rpm signal, or loss of signal pattern—No. 2 speed and timing sensor
91–08	32	Invalid throttle signal
91–13	28	Check throttle sensor calibration
102–03, 102–04	25	Boost pressure sensor circuit open or shorted
105–03, 105–04	38	Intake-manifold air temp. sensor circuit open or shorted
108–03, 108–04	26	Atmospheric pressure sensor circuit open or shorted
110–10	61	High coolant temp. warning
110–03, 110–04	27	Coolant temp. sensor circuit open or shorted
111–01	61	High coolant temperature
111–02	12	Coolant level sensor fault
164–00	17	Injector-actuation pressure out of range
164–02	15	Injector actuator pressure sensor erratic
164–03, 164–02	15	Injector actuator pressure circuit open or shorted
164–11	39	Injector actuation pressure system fault
168–02	51	Intermittent battery power to ECM
190–02, 190–11	34	Loss of engine rpm signal or no pattern to signal—No. 1 speed and timing sensor

90% of the engines go for 200,000 miles before the oil pan or a cylinder head must be removed for servicing. Some log half-million miles between overhauls. But with age and mileage problems do develop.

The HEUI injection system works as described for Caterpillar engines, with differences in detail and nomenclature. Table 6-4 translates Ford acronyms and describes sensor 7.3L function. "Supply" is the voltage going to the sensor and "signal" is the sensor output to the computer (ECM).

The fuel system underwent major changes including the addition of MAT sensors, replacement of the mechanical high-pressure oil pump (HPOP) with a mechanical unit, and more responsive turbochargers.

Table 6-4. Power Stroke sensor nomenclature and function

AP	Accelerator position sensor
	Along with other variables, the ECM uses the AP signal to determine injector oil-pressure, pulse width and timing. Supply 5.0V +/− 0.5V, signal 0.5–0.7V at idle, 4.5V at wide open throttle.
BARO	Barometric pressure sensor
	The ECM uses the BARO signal to adjust fuel timing, fuel quantity, and glow-plug on-time during high altitude operation. Supply 5.0V +/− 0.5V, signal 4.6V at sea level, decreasing at higher altitudes.
CMP	Camshaft position sensor
	The ECM uses the CMP signal to monitor engine rpm and tdc for Nos. 1 and 4 cylinders. This Hall-effect sensor generates a digital voltage signal of 12.0V high, 1.5V low.
DTC	Diagnostic trouble code
EBP	Exhaust back pressure sensor
	The ECM uses the EBP signal to control the exhaust pressure regulator (EPR). Supply 5.0V +/− 0.5V, signal 0.8–1.0V @ 14.7 psi at idle, increasing with engine speed and load, decreasing with altitude.
EOT	Engine oil temperature sensor
	The ECM uses the EOT signal to control glow-plug on time, EPR, idle rpm and fuel delivery and timing. Supply 5.0V +/− 0.5V, signal 4.37V @ 32°F, 1.37V @ 176°F, 0.96V @ 205°F.
EPR	Exhaust back pressure regulator
	The EPR operates hydraulically from oil taken off the turbocharger pedestal mount. When intake air temperature is less than 37°F (50°F on some models) and the engine oil temperature is less than 140°F (168°F on some models) the ECM energizes a solenoid valve that causes oil pressure to close a butterfly at the turbo exhaust outlet. The valve opens under load and as the engine warms.
GPC	Glow plug control
	The ECM energizes the GPC relay for 10–120 seconds depending on engine oil temperature and barometric pressure.
GPL	Glow plug light
	The ECM turns the "wait to start" lamp "on" for 1–10 seconds, depending on engine oil temperature and barometric pressure.
GPM	Glow plug monitor
	Used on 1997 and later California vehicles to report if glow plugs malfunction.
IAT	Intake air temperature sensor
	The ECM uses the IAT signal to regulate exhaust backpressure. Supply 5 V +/− 0.5V, signal 3.90V @ 32°F, 3.09V @ 68°F, 1.72V @ 122°F.
IPC	Injection pressure control sensor
	The ECM uses the IPC signal to match fuel delivery with load and to stabilize idle rpm. Signal 1.00V @ 580 psi, 3.22V @ 2520 psi.
IDM	Injector driver module
	The IDM receives cylinder-identification and fuel-demand signals from the ECM, and generates a 115 VDC, 10A signal for the appropriate injector, varying pulse width as required.

(Continued)

Table 6-4. **Power Stroke sensor nomenclature and function (*Continued*)**

IPR	Injection pressure regulator
	The ECM varies the duty cycle of the IPR to control oil pressure and the volume of fuel delivered. 0% = full return to sump (open valve), 100% = full flow to injectors (closed valve). Functioning is monitored by the Injector pressure control sensor.
IVS	Idle validation switch
	The IVS is an on-off switch that signals the ECM when the engine is idling. Signal 0V at idle, 12V off-idle.
MAP	Manifold absolute pressure sensor
	The MAP measures manifold pressure to limit turbo boost, optimize timing, and reduce over-fueling and smoke. Signal frequency: 111 Hz @ 14.7 psi, 130 Hz @ 20 psi, 167 Hz @ 30 psi.
MIL	Malfunction indicator lamp
	"Check engine" or "service engine" lamp that the ECM illuminates when certain system faults are present. Can also be used to retrieve trouble codes.
PCM	Powertrain control module
	PCM is the onboard computer that receives sensor inputs, calculates output signals to actuators, and generates diagnostic codes. The computer also controls transmission shift points, anti-skid braking, and other powertrain functions. Referred to in this text as the ECM.
PID	Parameter identification
	PID, or the data stream, is the sensor data read by scan tools.

7.3L Troubleshooting

Slow or no-start

Begin with a walk-around of the vehicle to determine that the fuel tank is at least a third full, batteries are charged, battery terminals are clean and secure, and the radiator is topped off. Check the crankcase and the HPOP reservoir oil levels. Use the recommended lube oil, such as 15W40 "black lid." The oil level should come to within 1 in. of the top of the reservoir.

Question the operator to learn as much as possible about the problem and conditions that led up to it. Did it appear suddenly? After repairs to the engine? What parts were replaced? Does the malfunction occur at normal engine temperature or only during cold starts?

Mechanics differ about what checks to make first. But the main bases to touch with the 7.3L are:

1. Replace the air filter and examine the exhaust system for restrictions. The exhaust butterfly valve, which cycles closed during cold starts, may stick shut. The backpressure regulator that controls this valve can also fail.
2. Retrieve active and historical trouble codes. P1111 means no codes are set and, so far as the ECM knows, things are normal.

3. Drain a sample of fuel from the filter while cranking. If water is present, drain the tank(s) and refill with clean fuel before proceeding. Inspect the water separator for oxidation damage and, if necessary, replace it.

4. Check the glow-plug relay, which is a high morality item and especially so in its original oval form. The newer, round-case relay interchanges with the older unit. The red cable, connected with fusible links to the starter relay, should have battery voltage at all times. If no voltage is present, check the fusible links and connections for opens. The heavy 10-gauge brown wire going to the glow plugs is hot when the relay is energized. One of the 18-gauge wires (usually red with a green tracer) is the signal wire, energized by the lube-oil sensor. This wire must have battery voltage for the relay to function. Depending upon the vehicle model, the "wait-to-start" lamp may signal when the glow plugs energize or count off seconds. Individual glow plugs should have a resistance of about 1 Ω cold and 2 Ω when cylinder temperatures stabilize.

5. Check fuel pressure when cranking for 15 seconds. If pressure is less than 50 psi, replace the fuel filter, remove any debris on the fuel screen and on the screen protecting the injection pressure regulator (IPR) deceleration orifice.

6. Check injector-oil pressure while cranking. A pressure of 500 psi is necessary to enable the injectors. Expect to see 960–1180 psi at high idle and 2500-plus psi during snap acceleration. If the computer senses that the IPR malfunctions, it holds oil pressure to a constant 725 psi. Trouble code 1280 means low IPR signal voltage, 1281 high signal voltage, and 1212 abnormally high (at least 1160 psi) oil pressure with the engine off. See "Low HPOP oil pressure" below.

7. While cranking, check for the presence of an rpm signal. A failed camshaft position sensor (CMP) can hold oil pressure below the 500-psi injector enabling threshold. See the "Camshaft position sensor" section below.

8. Check power at the injector solenoids with a scan tool. Zero pulse width on all injectors means a bad CMP sensor or injector driver module (IDM). Failure may trip trouble code 1298. The high power levels—10A at 115 VDC—put severe demands on the module. Modules on Econoline 7.3L vans built before 4/11/96 have problems with water intrusion.

   WARNING: Injector voltage can be lethal. Do not disconnect or otherwise tamper with injector wiring while the engine is running or the ignition key is ON.

9. Finally check for air in the fuel circuit by inserting a length of transparent tubing in the return fuel line.

Erratic idle

A frequent complaint is a rough idle, which can be sourced to the fuel system or to one or more weak cylinders. Follow this procedure:

1. Verify that the problem is not caused by a loose air-conditioner compressor bracket or another mechanical source.

2. Scan for trouble codes.
3. If the scanner permits, make a "buzz" test on the injectors. Listen for differences in the sound. A bad injector often buzzes at a lower frequency than the others.
4. Remove the valve covers and, with the engine idling, watch the oil flow out the injector spill ports. Replace any injector that passes noticeably less oil than the others.
5. Test cylinder compression.
6. As a last resort, replace the injectors.

Low HPOP oil pressure

If you have not already done so, check the oil level in the pump reservoir, which should come to within an inch of the top of the unit. If the engine has been idle for a long period, oil drains back out of the reservoir, making for hard starting.

Verify that there are no leaks at the line connections and around HEUI bodies. Remove the valve covers and look for cracks in the casting. Small leaks can be detected by pressurizing each gallery with an injector tester or a grease gun and a 3000-psi pressure gauge. Charge the gallery with oil at 1000 psi and wait several hours for any leaks to register on the gauge.

The next step in the search for causes of low oil pressure is to check the IPR, which functions like the Caterpillar IAPCV. If the 5V +/− 0.5V reference signal is present and the output circuit is not shorted or open, replace the regulator O-rings with parts available from Ford International or aftermarket sources. The DF6TZ- 9C977-AN Dipaco repair kit services IPRs used before and after engine SN 187099.

If O-rings do not solve the problem, the regulator should be tested by substitution of a known good unit. Torque to 35 ft/lb and do not use sealant that could clog the orifice on the threaded section.

Air in lube oil

Aeration often results in hard or no-starts, erratic idle, shutdowns on deceleration, and loss of rpm. CG4/SH oil helps control the problem. Reservoirs on 1994-model engines were specified as holding 12 quarts of oil. This specification was revised to 14 quarts, which means that the dipstick should be recalibrated or replaced with Ford PN F4TZ-6750-E.

In addition to its effect on performance, air in the oiling system causes rapid injector wear. Whenever an injector is changed, it is necessary to purge air from the associated cylinder bank. Disconnect the CMP to prevent the engine from starting.

- Mechanical fuel pumps (pre-1998 models)—disconnect the return line at the fuel pressure regulator on the fuel filter. Crank the engine in 15-second bursts, allowing ample time for the starter to cool, until a steady stream of fuel comes out the return line.
- Electric fuel pumps—crack a vent port on each cylinder head, and run the pump until air is purged.

If the head galleries have been drained, prefill the galleries with pressurized oil as described under "Low oil pressure."

Air in fuel circuit

Air leaks can cause rough idle, shutdowns on deceleration, and no-starts. The time required for air intrusion to make itself known depends upon the proximity of the leak to the transfer pump. For example, the engine might start and run several minutes before air entering from a fuel tank connection reaches the pump. Air entry closer to the pump has more immediate effects.

Multiple leak points—Schrader valves, bleed screws, fuel-line connections—make the filter/water separator a prime suspect. Use OEM clamps on hose connections, new seals, and dope screw threads with sealant. Filter bowl and filter drain-valve leaks can usually be repaired by substituting Viton O-rings, available from Guzzle's O-Rings (970-368-4455), for the Buna-N original parts. Surplus fuel from the injectors recycles back to the filter. The connection at the filter incorporates a check valve that closes when the engine stops. Should the check valve leak, air from the return line will be drawn into the filter as fuel in the canister cools and shrinks.

Injector driver module

The IDM delivers a low-side signal to control the timing, duration, and sequence of injector opening. The high-side delivers 115 VDC at 10A to the HEUI solenoids.

WARNING: Do not disconnect, attempt to measure or otherwise disturb IDM circuits while the engine is running or the key is ON. Voltages are lethal.

Camshaft position sensor

CMP sensors for mid-1994 to 1996 models are marked C96 or C 97; C-92 sensors were used from late-1997 to the end of production. Unless shimmed 0.010 in. off their mounting surfaces these OEM units can cause no-start, hard-start, erratic idle, and shutdowns on deceleration. Replacement sensors have been shortened and do not require shimming. The CMP for mid-1994 to 1996 models carries International part number C92 F7TZ-12K03-A and can be recognized by the tin-coated connectors. The latest C98-F7TZ-12K073-A version features gold-plated connectors, which gives one an idea of the problems associated with these units. While the newer CMP has better reliability than the previous type, its gold connectors present an electrolysis problem when mated with the tin connectors used on early 7.3L production. Replacement sensors cost about $100 from International. It's good insurance to carry a spare.

The CMP mounts on the front of the engine, adjacent to the camshaft, and is secured by two 10-mm bolts. Trouble code 0344 will be flagged if sensor response is intermittent; 0341 means that enough electrical noise has been detected to affect engine operation. Check connectors for loose, bent, or spread pins and clean grounds. Code 0340 indicates the absence of a sensor signal while cranking. In this case, the engine will not start.

7
CHAPTER

Cylinder heads and valves

The cylinder head acts as the backing plate for the head gasket and must, above all, be rigid. Iron is the preferred material, although automotive and other light-duty engines often are fitted with aluminum heads. Because of its good conductivity, aluminum assists cooling and eliminates the steep thermal gradients that iron is heir to. In other words, aluminum heads are less likely to develop local hot spots.

But, in common with most other structural materials, the weight savings of aluminum come at the cost of rigidity. Aluminum has a third the density of iron and a third the rigidity. To achieve the same level of stiffness, an aluminum head would have to weigh as much as the iron head it replaces. Good design practice and closely spaced head bolts ameliorate the problem, but aluminum heads are always less rigid than their cast-iron equivalents.

And while aluminum can be heat-treated to T6 or T7 hardness, few manufacturers bother, since the benefits of heat treating disappear at high temperatures. The light-metal alloys in general use exhibit a rapid loss of strength at temperatures above 200°C (392°F). Overheating an engine is almost guaranteed to warp the head. To further complicate matters, cast-iron has a lower rate of thermal expansion than aluminum. When bolted to an iron block, an aluminum head tends to bow upward, reducing the clamping force on the gasket. This phenomenon is discussed in the section that deals with head straightening.

The cylinder head forms an important part of the cooling system. The water jacket should be cast with large passages that resist clogging. Some designs employ diverters, or baffles, to direct the coolant stream to valve seats and other critical areas. Standard practice is to integrate block and head cooling, but several industrial engines separate the two so that a blown head gasket does not contaminate the crankcase oil with antifreeze.

Fasteners—capscrews or studs—should be arranged symmetrically insofar as rocker pedestals, valve ports, and injector and glow-plug bosses permit.

Combustion chamber types

A vast amount of development work was done in the 1920s and 30s on optimum combustion chamber shapes. The effort continues—with emphasis on control of emissions—but the main outlines of the problem and the alternatives appear to be firmly established.

The central problem is to provide a mechanism for mixing fuel with the air. The fuel spray must be dispersed throughout the cylinder so that all available oxygen takes part in the combustion process. As fuel droplets move away from the nozzle, the chances of complete combustion are progressively lessened. There is less oxygen available because some of the fuel has already ignited. In addition, the droplets shrink as they travel and successive layers of light hydrocarbons boil off.

Direct injection

Direct Injection (DI) chambers, also called open or undivided chambers, resemble those used in carbureted engines (Figure 7-1). These symmetrical chambers have small surface areas and, to reduce heat losses further, generally take the form of a cavity in the piston crown.

Until the advent of multiple-orifice, staged injectors and the high-pressure fuel systems necessary to support them, DI chambers compensated for poor fuel penetration by imparting energy to the air charge. Two mechanisms were involved: swirl and squish. The exit angle of the intake-valve seat imparted a spinning motion, or swirl, to incoming air stream. Squish was achieved by making the edges of the piston crown parallel to the chamber roof. As the piston neared top dead center (tdc), air trapped between these two faces "squished" inward, toward the piston cavity.

7-1 DI or open chamber.

Mixing was less than perfect, which resulted in ignition delay and rapid rises in cylinder pressure as the accumulated fuel charge exploded. As a result, DI was confined to stationary, heavy trucks and marine engines where noise, vibration, and exhaust smoke counted for less than fuel economy.

The breakthrough came in the late 1980s in the form of electronic injectors that sequenced fuel delivery to "soften" combustion and reduce ignition delay. Ultrahigh fuel pressures, coupled with orifice diameters as small as 0.12 mm, or twice the thickness of a human hair, atomize the fuel for better mixing. The fuel charge became the primary vehicle for mixing that, in some cases, enabled designers to eliminate the pumping losses associated with generating air turbulence. DI also opened the way for massive power increases. By substituting DI for the Ricardo Comet V cylinder head originally fitted, enlarging the bore, and adding a turbocharger, Volkswagen increased the power output of its signature four-cylinder engine by a factor of 3.4.

Indirect injection

Indirect injection (IDI) uses energy released by combustion to drive the fuel charge deeply into the air mass. The combustion chamber is divided into two sections with the larger chamber formed by the piston top and cylinder-head roof. Combustion begins in a smaller chamber, usually located over the piston. A narrow passage, often in the form of a venturi, connects the two chambers.

Fuel mixing, at least with low-pressure injectors, is enhanced since the fuel droplets exit the smaller chamber at very high velocities. Because peak pressures occur in the smaller chamber and do not act on the piston, IDI reduces noise and combustion roughness. The first diesel passenger cars would not have been commercially acceptable without IDI, and many light-duty engines continue to use this technology.

On the debit side, indirect injection imposes pumping losses since the piston must work to pressurize the small antechamber. In addition, this chamber acts as a heat sink, bleeding thermal energy that could be better employed in turning the crankshaft. For cold starting, the air charge must be heated with glow plugs. All of these chambers expel a jet of burning fuel that rebounds off the piston. A supercharger or turbocharger increases the temperature of the air charge and the vulnerability of the piston to meltdown.

Precombustion chamber

The precombustion chamber was first used on the Hornsby-Ackroyd low-compression oil engine and subsequently by Caterpillar, Deutz, and Mercedes-Benz, and other diesel manufacturers (Figure 7-2). The precombustion chamber, also known as a hot bulb because of its shape and absence of any provision for cooling, occupies 25–40% of the total swept volume.

As the piston approaches tdc, the injector opens to send a solid stream of fuel into the hot bulb. The charge ignites, bulb pressure rises, and a stream of burning fuel jets through the connecting channel and into the main chamber, where there is sufficient air to complete combustion. The smaller the channel, the greater the acceleration and, all things equal, the more complete the fuel-air mixing. These chambers employ pintle-type injectors.

Precombustion chamber

7-2 IDI in the form of a precombustion chamber.

Swirl chamber

The swirl, or turbulence, chamber is similar in appearance to the precombustion chamber, but functions differently (Figure 7-3). During compression, the disc-shaped or spherical antechamber imparts a circular motion to the air, which accelerates as the piston approaches tdc. The injector is timed to open at the peak of vortex speed. As the piston rounds tdc, combustion-induced pressure in the antechamber reverses the flow. A turbulent stream of burning fuel and superheated air exits the antechamber and rebounds off the piston to saturate the main chamber.

The swirl chamber was invented by Sir Harry Ricardo during the late 1920s and underwent numerous alterations during its long career. Except for Mercedes-Benz, most diesel cars and light commercial vehicles of the postwar era and for many years after employed the Ricardo Comet V chamber. These chambers are more economical than hot-bulb chambers, but are noisier.

The 6.25L engine used by General Motors for pickup trucks during the 1980s and early 90s demonstrates the tradeoffs implicit in combustion-chamber design. While the Ricardo chamber depends upon swirl for mixing, velocity is also important. Initially, these GM engines were set up with a small-diameter port between the main and swirl chambers. The pressure drop across the port generated velocity that, in conjunction with swirl, resulted in good air-fuel mixing and fuel economy. Near the end of the production run, GM acquiesced to customer demands for more power by enlarging the connecting port. More fuel could be passed, but thermal efficiency suffered. Fuel economy, which had approached 20 miles/gal, dropped to 14 or 15 mpg.

7-3 IDI in the form of a turbulence chamber, the most common of which were developed by Ricardo Ltd.

Energy cell

Examples of the energy cell, or Lanova divided chamber, can still be encountered in vintage Caterpillar tractor engines (Figure 7-4). The cell consists of a kidney-shaped main chamber, located over the piston, and a secondary chamber, or energy

Fuel injection   Combustion in energy cell

Fuel ignition   Combustion in main chamber

7-4 Hot bulb or Lanova IDI chamber is seen most often on older Cat engines.

cell, which is divided into two parts. The cell opens to a narrow throat, situated between the two lobes of the main chamber.

During the compression stroke about 10% of the air volume passes into the energy cell. The injector, mounted on the far side of the main chamber, delivers a solid jet of fuel aimed at the cell. A small percentage of the fuel shears off and remains in the main chamber and some collects in the cell chamber closest to the piston bore. But most of the fuel dead-ends in the outermost cell chamber.

Ignition occurs in this outer chamber. Unburned fuel and hot gases accelerate as they pass through the venturi between the two cell chambers and enter the main chamber. The rounded walls of the main chamber impart a swirl to the charge to promote better mixing.

Valve configuration

Modern engines employ overhead valves (ohv), although side-valve (sv) engines are occasionally encountered. Mounting the valves over, rather than alongside, the piston makes for a cleaner, less heat-absorbent combustion chamber and permits larger valve diameters.

Overhead valves are generally driven from a block-mounted camshaft through pushrods and rocker arms (Figure 7-5). High-performance auto engines and commercial engines with unit injectors employ a single overhead camshaft (ohc) that acts on the valves through rockers. The most recent Mercedes-Benz automotive diesel breaks with tradition by using double overhead camshafts (dohc).

For good volumetric efficiency, valves should be as large as possible. Some American trucks and many European passenger cars have four valves per cylinder to improve breathing and power output. With a pent-roof chamber, four valves can provide a flow area of about a third of the area of the piston crown. In contrast, the flow area of two parallel valves is only a fifth of the crown area.

When purchasing aftermarket valves, it is useful to know that the SAE classifies valve steels into four groups:

- NV low-alloy steels for intake valves,
- HNV high-alloy steels for intake valves,
- EV austenitic steels for exhaust valves; and
- HEV high-strength alloy for exhaust valves.

When upgrading intake valves, one would do well to specify SAE-rated EV8 21-4N austenitic steel that contains 21% chromium and 3.75% nickel, and tolerates temperatures of 1600°F (871°C). Inconel 751, which the SAE classifies as HEV3, is preferred for exhaust valves because of its hot strength. This nickel-based alloy contains 16% chromium and 3% titanium.

Regardless of the steel used, long-lived exhaust valves use sodium as a medium of heat exchange. When heated, sodium turns liquid and transfers about 40% of the heat load to the valve guide. In contrast, solid-steel valves transfer only 25% of the heat to the guide, the rest of it going through the seat. Because sodium combines

1 Locknut
2 Rocker shaft
3 Adjusting screw
4 Pushrod
5 Nozzle assembly
6 Glow plug cable
7 Glow plug
8 Combustion chamber
9 Rocker cover
10 Valve rocker
11 Spring seat
12 Split collar
13 Stem seal
14 Valve spring
15 Valve
16 Valve seat

1 Locknut
2 Valve rocker
3 Adjusting screw
4 Rocker shaft
5 Spring seat
6 Split collar
7 Stem seal
8 Valve spring
9 Valve
10 Valve seat
11 Pushrod
12 Valve lifter
13 Camshaft

7-5 Typical ohv valve train, driven from the camshaft via pushrods and rocker arms. Marine Engine Div., Chrysler Corp.

Full tulip exhaust

Semitulip exhaust

Tulip intake

Mushroom intake

Mushroom exhaust

7-6 Tulip and mushroom valves were used on vintage engines. Contemporary practice is to flatten the valve crown to reduce heating.

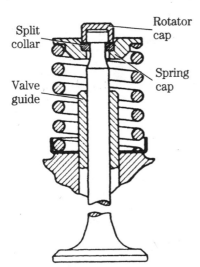

Split collar

Rotator cap

Valve guide

Spring cap

7-7 The GM Bedford valve assembly includes rotator cap (Rotcap), split collar keepers, spring cap, and spring.

explosively with water, these valves should be treated with respect. Discarded valves should be clearly marked as containing sodium.

Valve inserts, or seats, are an interference fit in the cylinder head. In addition to forming a gas seal with the valve face, the seats provide a major path for heat transfer from the valve face. Consequently, the seats are adjacent to fins on air-cooled engines or surrounded by water on liquid-cooled types. Sintered-iron alloys are often used for intake seats and nickel-steel for the exhausts.

While valves are forced open mechanically by the camshaft, spring pressure seats them. As the springs weaken in service, it is possible for the valve train to "lead" the cam. The acceleration imposed during lift times the reciprocating weight of the train (cam followers, pushrods, half of the mass of the rocker arms, and valves) over-comes spring tension and the valves float open. A similar problem occurs during seating: insufficient spring tension permits valves to bounce several times before coming to rest. These effects become more severe as engine speeds increase.

The angle of the faces varies with many manufacturers using a 30° angle on the intakes and 45° on the exhausts. The steeper angle imposes a flow restriction, but generates the higher seating forces needed for exhaust valves. Valve guide material ranges from chilled iron to manganese or silicon bronze.

Before you begin

Some repairs, such as head-gasket replacement, are normally done on site. Other operations require the assistance of a machinist. But the mechanic needs to understand enough about the compromises involved in returning worn parts to service to be able to evaluate the machinist's work. And, as far as possible, the mechanic should keep current with the technology, such as diamond-like carbon (DLC), titanium nitride (TnN), and chromium nitride (CrN) anti-friction coatings, and isotropic polishing. The latter involves the use of irregularly shaped ceramic pellets in a vibratory machine. The process, which takes about ten hours to complete, extends valve and valve-train life by removing stress risers and microscopic peaks.

Replacement parts can be purchased from the original equipment manufacturer (OEM) or, for the more popular engines from the aftermarket. Some aftermarket parts are equal to those supplied to the OEM and may, in fact, be the same parts. Others are not so good in ways that might not be apparent until the engine runs a few hundred hours.

Counterfeit parts, disguised as OEM parts, are widely marketed and sometimes by authorized factory dealers. Be suspicious of a part that costs considerably less than it should or that differs from the original in finish, code number, or packaging.

Diagnosis

Head-related problems usually involve loss of compression in one or two cylinders, a condition that is signaled by a ragged idle, by increased fuel consumption, and in some instances, by exhaust smoke. An exhaust temperature gauge, with switch-controlled thermocouples on each header, will give early warning. Exhaust from weak cylinders will be cooler than the norm. You can cross-check by disabling one injector at a time while the engine ticks over at idle. A cylinder that causes less

7-8 Cylinder compressor gauge and adapter. Peugeot.

of an rpm drop than the others does not carry its share of the load. (See Chap. 4 for additional information about this procedure.)

At this point, you can check the injectors (the usual suspects) or else go to the heart of the matter with a cranking compression test (Figure 7-8). As detailed in the diagnostics chapter, we are looking for cylinders with dramatically (at least 20%) lower compression than the average of the others. If the weak cylinder is flanked by healthy cylinders, the problem is either valve- or head-gasket related; or very low compression in an adjacent cylinder points to gasket failure. Abnormally high readings on all cylinders indicate heavy carbon accumulations, a condition that might be accompanied by high combustion pressures and noise. The next step is to make a cylinder leak-down test, which will distinguish between valve- or head-gasket failure.

While most compression leaks bleed into adjacent cylinders or across the fire deck to the atmosphere, it is possible for a leak path to open into water jacket. The engine might seem healthy enough, but overheat within a few minutes of start-up. Coolant in the header tank might appear agitated and might spew violently with the cap removed. A cooling system pressure test will verify the existence of a leak, which can be localized with a cylinder leak-down test. However, the leak-down test cannot distinguish between cracks in the casting and a blown gasket.

Fortunately, it is rare for an engine that has not suffered catastrophic overheating to leak coolant into the oil sump, where it can be detected visually or, in lesser amounts, by a spectrographic analysis. Likely sources are casting cracks, cracked (wet-type) cylinder liners, and liner-base gasket leaks.

The cylinder head casting, like the fluid end of a high-pressure pump, will eventually fail. After a large, but finite, number of pressure cycles, the metal crystallizes and breaks. Owners of obsolete engines for which parts are no longer available would do well to keep a spare head casting on hand. Even so, most cylinder heads fail early, long before design life has been realized, because of abnormally high combustion pressure and temperature.

Combustion pressure and heat can be controlled by routine injector service (dribbling injectors load the cylinders with fuel), attention to timing, and conservative pump settings. Cooling system maintenance usually stops when the temperature gauge needle remains on the right of center. But local overheating is as critical as radiator temperature and is rarely addressed. According to engineers at Detroit Diesel, ¼ in. of scale in the water jacket is the thermal equivalent of 4 in. of cast iron. Eroded coolant deflectors and rounded-off water pump impeller blades can also produce local overheating, which will not register as a rise in header-tank temperature.

Local overheating on air-cooled engines can usually be traced to dirty, grease-clogged fins or to loose shrouding.

Gasket life can be extended by doing what can be reasonably done to minimize potential leak paths. As explained below, some imperfection of the head and fire deck seating surfaces must, as a practical matter, be tolerated. Fasteners should be torqued down to specifications and in the suggested torque sequence, which varies between engine makes and models. Asbestos gaskets, without wire or elastomer reinforcements, "take a set" and must be periodically retightened. Newer, reinforced gaskets hold initial torque, but proper diesel maintenance entails retightening head bolts periodically.

Disassembly

The drill varies between makes and models, and is described in the manufacturer's manual. Here, I merely wish to add some general information, which might not be included in the factory literature.

It is good and sometimes necessary practice to align the timing marks before the head is dismantled. Bar crankshaft over—in its normal direction of travel—to tdc on No. 1 cylinder compression stroke. Top dead center will be referenced on the harmonic balancer or flywheel; the compression stroke will be signaled by closed intake and exhaust valves on No. 1 cylinder. In most cases, the logic of timing marks on overhead cam and unit injector engines will be obvious; when it is not, as for example, when camshaft timing indexes a particular link of the drive chain, make careful notes. A timing error on assembly can cost a set of valves.

Experienced mechanics do not disassemble more than is necessary. Normally you will remove both manifolds, unit injector rocker mechanisms, and whatever hardware blocks access to the head bolts. Try to remove components in large bites, as assemblies, by lifting the intake manifold with turbocharger intact, removing the shaft-type rocker arms at the shaft hold-down bolts, and so on.

Miscellaneous hardware should remain attached, unless the head will be sent out for machine work. In this case, it should be stripped down to the valves and injector tubes. Otherwise, the head might be returned with parts and fasteners missing.

Note: Injector tubes on some DI heads extend beyond the head parting surface and, unless removed, might be damaged in handling.

Examine each part and fastener as it comes off. If disassembly is extensive, time will be saved by storing the components and associated fasteners in an orderly fashion. You might wish to use plastic baggies, labeled with a Sharpie pen or Marks-A-Lot for this purpose. Keep the old gaskets for comparison with the replacements.

Intake valve
rocker arm

Injector
rocker arm

Exhaust valve
rocker arm

Roller
follower

Camshaft

7-9 Detroit Diesel Series 60 valves and injectors actuate from a single overhead camshaft.

Rocker assemblies

Two rocker arm configurations are used on OHV engines. Commercial and automotive engines developed from industrial engines generally pivot the rockers on single or double shafts, as illustrated several times in this chapter, including Figure 7-6 and, as applied to ohc engines, in Figures 7-9 and 7-10. The earlier drawing is the most typical, because a single rocker drives each intake and exhaust valve. Figures 7-9 and 7-10 illustrate Detroit Diesel solutions to the problem of driving two valves from the same rocker arm. Regardless of the rocker configuration, the hollow pivot shaft doubles as an oil gallery, distributing oil to the rocker bushings through radially drilled ports. Shaft-mounted rockers are steel forgings, case-hardened on the valve end and generally include provision for valve lash adjustment.

7-10 Detroit Diesel dual valve actuating mechanism, used on the firm's two-cycle engines.

These rockers are detached from the head as an assembly with the shaft while the pushrods are still engaged. Note the lay of the shaft and rockers, and if no identification marks are present, tag the forward end of the shaft as an assembly reference. Loosen the hold-down bolts slowly, a half-turn or so at a time following the factory-recommended breakout sequence. If no information is provided on this point, work from the center bolt outward. This procedure will distribute valve spring forces over the length of the shaft.

Set the rocker arm assembly aside and remove the pushrods, racking them in the order of removal with the cam-ends down. A length of four-by-four, drilled to accept the pushrods and with the front of the engine clearly marked, makes an inexpensive rack.

Although the task is formidable on a large engine, the rocker arms, together with spacers, locating springs, and wave washers, should be completely disassembled for cleaning and inspection. Critical areas are:

- *Adjusting screws*—check the thread fit and screw tips. Screws are case-hardened: once the carburized "skin" is penetrated, it will be impossible to keep the valves adjusted.
- *Rocker tips*—with proper equipment, worn tips can usually be recontoured, quieting the engine.
- *Bushings*—wear concentrates on the engine side of the bushing and, when severe, is accompanied with severe scoring, which almost always involves the shaft. Clearance between the bushing and an unworn part of the shaft should be on the order of 0.002 in. Replacement bushings are generally available from the original equipment manufacture (OEM) or aftermarket. The old bushing is driven out with a suitable punch, and the replacement pressed into place and reamed to finish size. Aligning the bushing oil port with the rocker arm port is, of course, critical.
- *Shaft*—inspect the surface finish, mike bearing diameters, and carefully clean the shaft inner diameter (ID), clearing the oil ports with a drill bit. Do the same for the oil supply circuit. Rocker arms are remote from the pump, and lubrication is problematic.

Small engines in general and automotive plants derived from SI engines use pedestal-type rockers, of the type illustrated in Figure 7-11. This technology, pioneered by Chevrolet in 1955, represents a considerable cost saving because the pivots compensate for dimensional inaccuracies and the rockers are steel stampings. Most examples lubricate through hollow pushrods. If rockers are removed, it is vital that they be assembled as originally found, together with fulcrum pieces and hold-down hardware. Rack the pushrods as described in the previous paragraph.

The locknuts that secure the rockers to their studs should be renewed whenever the head is serviced. Other critical items are:

- *Studs*—check for thread wear, nicks, distortion, and separation from the head. As far as I am aware, all diesel pedestal-type rocker studs thread into the cylinder head.

Adjust valve
clearance here

Valve
spring
retainer

Valve
spring

Oil seal
(intake
valve only)

Rocker
arm

Valve
push
rod

A932

Short
sleeve
down

7-11 Pedestal-type rocker arm, pivots on an adjustable
fulcrum for lash adjustment. Onan

- *Rocker pivots*—the rocker pivots on a ball or a cylindrical bearing, secured by
the stud nut, and known as the fulcrum seat (Figure 7-12A). Reject the rocker,
if either part is discolored, scored, or heat checked. How much wear is permissible on the rocker pivot is a judgment call.
- *Fasteners*—replace locknuts if nut threads show low resistance to turning, or
for the considerable insurance value. Replace stud nuts if faces exhibit fractures
(Figure 7-12B).
- *Rocker tips*—look for evidence of impact damage that could point to a failed
hydraulic lifter and possible valve tip, valve guide, or pushrod damage. For want of
a better rule, replace the rocker when tip wear is severe enough to hang a fingernail.

Pushrods

Inspect the push rod for wear on the tips and for bends. The best way to determine
trueness is to roll the rods on a machined surface or a piece of optically flat plate glass.

Valve lifters

Valve lifters, or tappets, are serviced at this time when the lifters are driven from
an overhead camshaft or when battered rocker arm tips indicate the need. It should be
mentioned that GM 350 hydraulic lifters must be bled down before the cylinder head
is reinstalled. Factory manuals recommend that lifters be collapsed twice, the second
time 45 minutes after the first. According to technicians familiar with these engines,
the factory is not kidding.

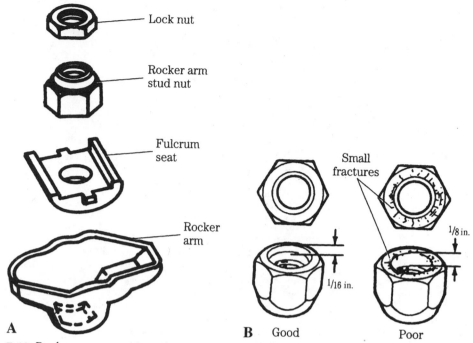

Lock nut

Rocker arm
stud nut

Fulcrum
seat

Rocker
arm

Small
fractures

1/8 in.

1/16 in.

A

B   Good   Poor

7-12 Rocker arm assembly and nomenclature (A); stud nut wear pattern (B). Ford Motor Co.

Figure 7-13 illustrates the roller tappet used on GM two-cycle engines. This part is subject to severe forces and, in the typical applications, experiences fairly high wear rates. Inspect the roller for scuffing, flat spots, and ease of rotation. Damage to the roller OD almost always is mirrored on the camshaft.

.010 in. total (max)
Diametric
clearance

.015 in.–0.23 in. total
Side
clearance

7-13 Roller tappet, with wear limits. General Motors Corp.

7-14 Light-duty engines such as the Ford 2.3L turbo may employ a cogged (or Gilmer) belt for camshaft drive. Gilmer belts might fail without obvious warning and should be replaced on a rigid engine-hours or mileage schedule. Belts must be tensioned as described by the engine maker and should not be subjected to reverse rotation. Baring the engine backwards can shear or severely damage the teeth.

Roller and pin replacement offers no challenge, but lubrication is critical. During the first few seconds of operation, the only lubrications the follower receives is what you provide. Engine-supplied lube oil is slow to find its way between the roller and pin.

Ensure proper lubrication by removing the preservative from new parts with Cindol 1705 and clean used parts with the same product. Just before installation, soak the followers in a bath of warm (100°–125°F) Cindol. Turn the rollers to release trapped air.

Install with the oil port at the bottom of the follower pointed away from the valves. There should be 0.005-in. clearance between the follower legs and guide. The easiest way to make this adjustment is to loosen the bolts slightly and tap the ends of the guide with a brass drift. Bolts should be torqued to 12–15 ft-lb.

Overhead camshafts

Disengage the drive chain or belt (Figure 7-14). Camshaft mounting provisions vary; some ride in split bearings and are lifted vertically, others slip into full-circle bearings and might require a special tool to open the valves temporarily for lobe clearance during camshaft withdrawal. Split bearing journals must be assembled exactly as originally found. Make certain that bearing caps are clearly marked for number and orientation. Loosen the caps one at a time, working progressively from the center

Fuel injector

Precombustion chamber

Precup

7-15 Peek-a-boo head bolts hide under pipe plugs on the infamous GM 350 engines. Upon assembly, plug threads should be coated with sealant. Fel-Pro, Inc.

cap out to the ends of the shaft. (Center, first cap right of center, first cap left of center, second cap right of center, and so on.)

Head bolts

Head bolts should now be accessible, but not always visible. Olds 350 engines hide three of the bolts under pipe plugs (Figure 7-15); some Japanese engines secure the timing cover to the head with small-diameter bolts that, more often than not, are submerged in a pool of oil.

The practice of using an impact wrench on head bolts should be discouraged. A far better procedure and one that must be used on aluminum engines is to loosen the bolts by hand in three stages and in the pattern suggested by the manufacturer. Make careful note of variations in bolt length and be alert for the presence of sealant on the threads. Sealant means that the bolt bottoms into the water jacket, a weight-saving technique inherited from SI engines.

Note: To prevent warpage, the head must be cold before the bolts are loosened.

Clean the bolts and examine carefully for pulled threads, cracks (usually under the heads), bends, and signs of bottoming. Engines have come off the line with short bolt holes.

GM and several other manufacturers use torque-to-yield bolts, most of which are throwaway items. When this is the case, new bolts should be included as part of the gasket set.

Lifting

Large cylinder heads require a lifting tackle and proper attachment hardware (Figure 7-16); heads small enough to be manhandled might need a sharp blow with a rubber mallet to break the gasket seal. Lift vertically to clear the alignment pins, which are almost always present.

Mount the head in a holding fixture (Figure 7-17), or lacking that, support it on wood blocks.

Valves

Valves are removed by compressing the springs just far enough to disengage the keepers, or valve locks (Figure 7-18). It is good practice to replace valve rotators,

7-16 Lift brackets should be at the approximate center of gravity of the casting, so that the head lifts vertically off its alignment pins. Detroit Diesel

7-17 Cylinder head holding fixture. Marine Engine Div., Chrysler Corp.

shown in Figure 7-7. Seals (Figure 7-19) should also be renewed. Specify high temperature Viton for the seal material.

If all is right, the valves will drop out of their guides from their own weight. The usual cause of valve bind is a mushroomed tip, which can be dressed smooth with a stone and which means that the engine has been operating with excessive lash. The associated camshaft lobe might be damaged. A bent valve suggests guide seizure or piston collision.

Cleaning

Cleaning techniques depend on the available facilities. In the field, cleanup usually consists of washing the parts in kerosene or diesel fuel. Gasket fragments can be scraped off with a dull knife (a linoleum knife with the blade ground square to the handle is an ideal tool). The work can be speeded up by using one of the aerosol preparations that promise to dissolve gaskets. It is not a good practice to use a wire brush on head and cooling system gaskets that may contain asbestos. Some mechanics remove the sealant used in lieu of gaskets on valve covers and oil pans with 3M Scotch-Brite Surface Conditioning Discs, mounted on a high-speed die grinder. The coarse pad (3M 07450) is used on steel surfaces, the medium (3M 07451) on aluminum. However, it is not a good idea to grind on an unknown chemical substance.

7-18 Lever-type valve spring compressor.

Valve top retainer

Intake and exhaust
valve spring collar

Intake and exhaust
valve spring seat

Intake and exhaust valve

Valve step seal

Valve spring

7-19 Yanmar valves employ umbrella-type seals.

Carbon responds to a dull knife and an end-cutting wire wheel. Clean the piston tops, rotating the crank as necessary. Overhead camshaft drive chains will foul if the crankshaft is turned, and one cannot pretend to do much by way of piston cleaning on these engines.

Machine shops and large-scale repair depots employ less labor-intensive methods. Some shops still use chlorinated hydrocarbons (such as perchloroethylene and trichloroethylene) for degreasing, although the toxicity of these products has limited their application. A peculiar side effect of trichloroethylene (TCE) exposure is "degreaser's flush." After several weeks of contact with the solvent, consumption of alcohol will raise large red welts on the hapless degreaser's face.

Once the head (and other major castings) are degreased, ferrous parts are traditionally "hot-tanked" in a caustic solution, heated nearly to the boiling point. Caustic will remove most carbon, paint, and water-jacket scale. Parts are then flushed with fresh water and dried. In the past, some field mechanics soaked iron heads in a mild solution of oxalic acid to remove scale and corrosion from the coolant passages.

Caustic and other chemical cleaners pose environmental hazards and generate an open-ended liability problem. The Environmental Protection Agency holds the producer of waste responsible for its ultimate disposition. This responsibility cannot be circumvented by contract; if a shop contracts for caustic to be transported to a hazardous waste site, and the material ends up on a country road somewhere, the shop is liable.

Consequently, other cleaning technologies have been developed. Pollution Control Products is perhaps the best-known manufacturer of cleaning furnaces and claims to have more than 1500 units in service. These devices burn natural gas, propane, or No. 2 fuel oil at rates of up to 300,000 Btu/hour to produce temperatures of up to 800°F. (Somewhat lower temperatures are recommended for cylinder heads and blocks.) An afterburner consumes the smoke effectively enough to meet EPA emission, OSHA workplace safety, and most local fire codes.

7-20 Walker Peenmatic shot blaster adds an "as-new" finish used casting.

Such furnaces significantly reduce the liability associated with handling and disposal of hazardous materials, but are not, in themselves, the complete answer to parts stripping. Most scale flakes off, but some carbon, calcified gasket material, and paint might remain after cleaning.

Final cleanup requires a shot blaster such as the Walker Peenimpac machine illustrated in Figure 7-20. Parts to be cleaned are placed on a turntable inside the machine and bombarded with high-velocity shot. Shot size and composition determine the surface finish; small diameter steel shot gives aluminum castings a mat finish, larger diameter steel shot dresses iron castings to an as-poured finish. Delicate parts are cleaned with glass beads.

Head casting

Make a careful examination of the parting surfaces on both the head and block, looking for fret marks, highly polished areas, erosion around water-jacket ports, and scores that would compromise the gasket seal.

The next step is to determine the degree of head distortion, using a machinist's straightedge and feeler gauges, as illustrated in Figure 7-21. Distortion limits vary with the application and range from as little as 0.003–0.008 in. or so. In theory, the block deck has the same importance as the head deck; in practice, a warped block would not be welcome news when head work was all that was contemplated, and most mechanics let sleeping dogs lie.

Straightedge

Feeler gauge

7-21 Head warp should be checked in three planes: diagonally as shown, longitudinally, and transversely with particular attention to areas between combustion chambers. Ford Motor Co.

However head bolt holes deserve special attention. Be alert for:

- *Stripped threads*—strip-outs can be repaired with Heli-Coil inserts. It is rare to find more than one bolt hole stripped; if the problem is endemic, check with the manufacturer's technical representative on the advisability of using multiple inserts. At least one manufacturer restricts the number of Heli-Coils per engine and per cylinder.
- *Pulled threads*—this condition, characterized by slight eruption in the block metal adjacent to the bolt holes, can be cured by chamfering the uppermost threads with an oversized drill bit or countersink. Limit the depth of cut to one or two threads.
- *Dirty threads*—chase head bolt and other critical threads with the appropriate tap. In theory, one should use a bottoming tap, recognized by its straight profile and squared tip. In practice, one will be lucky to find any tap for certain metric head bolt threads, which are pitched differently than the run of International Standards Organization (ISO) fasteners. Also realize that "bargain" taps can, because of dimensional inaccuracies, do more harm than good. Blow out any coolant that has spilled into the bolt holes with compressed air, protecting your eyes from the debris.

Head resurfacing

Minor surface flaws and moderate distortion can usually be corrected by resurfacing, or "milling." However, there are limits to how much metal can be safely removed from either the head or the block. These limits are imposed by the need to maintain piston-to-valve clearance and, on overhead cam engines, restraints imposed by the valve actuating gear. The last point needs some amplification. Reducing the thickness of the head retards valve timing when the camshaft receives power through a chain or belt. Retarded valve timing shifts the torque curve higher on the rpm band. The power will still be there, but it will be later in coming. The effect of lowering a gear-driven camshaft is less ambiguous; the gears converge and ultimately jam.

Fire deck spacer plates are available for some engines to minimize these effects, and ferrous heads have been salvaged by metal spraying. These options are worth investigating, but one is usually better off following factory recommendations for head resurfacing, as in other matters.

The minimum head thickness specification, expressed either as a direct measurement between the fire deck and some prominent feature on the top of the casting or as the amount of material that can be safely removed from the head and block. Detroit Diesel allows 0.020 in. on four-cycle cylinder heads and a total of 0.030 in. on both the heads and block. Other manufacturers are not so generous, especially on light- and medium-duty engines. For example, the head thickness on Navistar 6.9L engines (measured between the fire deck and valve cover rail) must be maintained at between 4.795 and 4.805 in., a figure that, when manufacturing tolerances are factored in, practically eliminates the possibility of resurfacing. As delivered, the exhaust valve might come within 0.009 in. of the piston crown, and the piston crown clears the roof of the combustion chamber by 0.025 in. Admittedly, these are minimum specifications, but it is difficult to believe that any 6.9L can afford to lose much fire deck metal. Navistar's 9.0L engine, the Ford 2.2L, and the Volkswagen 1.6L engines simply cannot be resurfaced and remain within factory guidelines. Thicker-than-stock head gaskets are available for the VW, but are intended merely to compensate for variations in piston protrusion above the fire deck.

The cardinal rule of this and other machining operations is to remove as little metal as possible, while staying within factory limits. Minor imperfections, such as gasket frets and corrosion on the edges of water jacket holes, should be brazed slightly overflush before the head is milled.

Precombustion chambers, or precups, are pressed into the head or Caterpillar engines, threaded in. In some cases, precups can remain installed during resurfacing; others must be removed and (usually) machined in a separate operation (Figure 7-22) When it is necessary to replace injector tubes, the work is done after the head has been resurfaced and is always followed by a pressure test.

Note: Ceramic precups are integral components, serviced by replacing the head.

Most shops rely on a blanchard grinder, such as the one shown in Figure 7-23 for routine resurfacing. Heavier cuts of 0.015 in. or more call for a rotary broach, known in the trade as a "mill." When set up correctly, a mill will give better accuracy than is obtainable with a grinder. Some shops, especially those in production work, use a movable belt grinder. These machines are relatively inexpensive, require zero set-up time, and produce an unsurpassed finish. However, current belt grinders, which support the workpiece on a rubber platen, are less accurate than blanchard grinders.

Typically, iron heads like a dead smooth surface finish in the range of 60–75 rms (root mean square). Some machinists believe that a rougher finish provides the requisite "tooth" for gasket purchase, although there is little evidence to support the contention.

Crack detection

Cylinder heads should be crack tested before and after resurfacing. The apparatus used for ferrous parts generates a powerful magnetic field that passes through the part under test. Cracks and other discontinuities at right angles to the field become polarized and reveal their presence by attracting iron filings. The magnetic particle test,

- Heat insulator
- Precombustion chamber (rear)
- Packing
- Pin
- Precombustion chamber (front)
- Packing

7-22 Precombustion chamber and gaskets. An installed precup is shown in Figure 7-15. Yanmar Diesel Engine Co. Ltd.

7-23 A blanchard grinder is used for light head milling, manifold milling, and other jobs where some loss of precision can be tolerated.

7-24 Penetrant dye detects cracks in nonferrous parts, a category that includes Valve heads. Yanmar Diesel Engine Co. Ltd.

known generically by the trade name Magnaflux, is useful within its limits. It cannot detect subsurface flaws, nor does it work on nonferrous metals. But fatigue and thermal cracks always start at the surface, and will be seen.

Nonferrous parts are inspected with a penetrant dye (Figure 7-24). A special dye is sprayed on the part, the excess is wiped off, and the part is treated with a developer that draws the dye to the surface, outlining the cracks. In general, penetrant dye is considered less accurate than Magnaflux, but, short of x-ray, it remains the best method available for detecting flaws in aluminum and other nonmagnetic parts.

Neither of these detection methods discriminates between critical and superficial cracks. The cylinder head might be fractured in a dozen places and still be serviceable. But cracks that emanate from a pressure regime—that is, from the combustion chamber, cylinder bore, water jacket, or oil circuits—require attention.

Mack and a handful of other manufacturers build surface discontinuities into the roofs of the combustion chambers, which appear as cracks under Magnafluxing, but which are intended to stop crack propagation. These built-in "flaws" run true and straight, in contrast to the meandering paths followed by the genuine article. In general, cracks that are less than a $1/2$ in. long and do not extend into the valve seat area might be less serious than they appear. Short cracks radiating out from precombustion chamber orifices can also be disregarded.

Crack repairs

Assuming that both ends of the crack are visible, it is normally possible to salvage an iron head by gas welding. For best results, the casting should be preheated to 1200°F. Skilled TIG practitioners can do the same for aluminum heads.

Some shops prefer to use one of several patented "cold stitching" processes on iron castings. The technician drills holes at each end of the crack (to block further propagation) and hammers soft iron plugs into the void, which are then ground flush. The plugs must not be allowed to obstruct coolant passages.

Metal spraying is demonstrated to have utility, but rarely practiced.

Test strips

7-25 The water jacket should be sealed as shown, pressurized to about 30 psi, and immersed in hot water. Bubbles indicate the presence of cracks.

Pressure testing

Figure 7-25 illustrates a Detroit Diesel cylinder head, partially dressed out for pressure testing. At this point, most shops would introduce high-pressure water into the jacket and look for leaks, which might take the form of barely perceptible seepage. Detroit Diesel suggests that air be used as the working fluid. Their approach calls for pressurizing the head to 30 psi and immersing the casting in hot (200°F) water for 30 minutes. Leaks appear as bubbles.

Straightening

Aluminum heads usually carry the camshaft and are often mounted to a cast-iron block. The marriage is barely compatible. Aluminum has a thermal coefficient of expansion four times greater than that of iron. Even at normal temperatures, the head casting "creeps," with most of the movement occurring in the long axis (Figure 7-26). Pinned at its ends by bolts, the head bows upward—sometimes as much as 0.080 in. The camshaft cannot tolerate misalignments of such magnitude and, if it does not bend, binds in the center bearings.

Such heads are best straightened by stress relieving. The process can be summarized as follows: the head is bolted to a heavy steel plate, which has been drilled and tapped to accommodate the center head bolts. Shims, approximately half the thickness of the bow, are placed under the ends of the head, and the center bolts are lightly run down. Four or five hours of heat soak, followed by a slow cool down, usually restores the deck to within 0.010 in. of true. Camshaft bearings are less amenable to this treatment, and will require line boring or honing.

Corrosion can be a serious problem for aluminum heads, transforming the water jacket into something resembling papier-mâché. Upon investigation, one often finds that the grounding strap—the pleated ribbon cable connecting the head to the firewall—was not installed.

Aluminum

Cast iron

Bi-metal engine expansion

7-26 Thermal expansion of an aluminum head is about four times greater than for a cast iron block. Something must give. Fel-Pro, Inc.

Valve guides

Rocker-arm geometry generates a side force, tilting the valves outward and wearing away the upper and lower ends of the guides. The loss of a sharp edge at the lower end of the guides encourages carbon buildup and accelerates stem wear; the bellmouth at the upper end catches oil, which then enters the cylinder. Figure 7-27 illustrates a split ball gauge used to determine guide ID.

Nearly all engines employ replaceable guides or, if lacking that, have enough "meat" in the casting to accept replaceable guides (Figure 7-28). A Pep replacement guide for 0.375-in. valve stem measures 0.502 in. on the OD. The BMW 2.4L engine is one of the few for which replacement guides are not available. However, the integral (i.e., block metal) guides can be reamed to accept valves with oversized stems.

Old guides drive out with a punch, and new guides install with a driver sized to pilot on the guide ID (Figure 7-28). Cast-iron heads can be worked cold, but a careful technician will heat aluminum heads so that the guide bores do not gall. Engine manufacturers seem to prefer perlitic cast-iron or iron-alloy guides although many machinists claim phosphor bronze has better wearing qualities. Whatever the material, replacement guides rarely are concentric with the valve seat, and some corrective machine work is almost always in order. Stem-to-guide clearances vary with engine type and

7-27 A split-ball gauge, used to determine guide wear. Ford Motor Co.

7-28 Valve guides are installed to a specified depth by means of a factory-supplied driver. Yanmar Diesel Engine Co. Ltd.

service; light- and medium-duty engines will remain oil-tight longer with a 0.0015-in. clearance. Heavy-duty engines, which run for long periods at half-rated power, need to be set up looser—as much as 0.005 in. when sodium-cooled valves are fitted.

Valve service

Stem and guide problems can be the result of wear or carbon and gum accumulations that hold the valve open against spring pressure. These deposits are caused by the wrong type of lubricant, ethylene glycol leakage into the sump, and low coolant temperatures (below 160°F). Low temperatures are usually the result of long periods at idle and aggravate any mismatch between fuel characteristics and engine demands. Fuel that burns cleanly at normal temperatures can gum the guide and carbon over the valve heads when the engine runs consistently cool. Sticking can also be caused by bent stems.

Premature valve burning, which in extreme cases torches out segments of valve face and seat, has one cause: excessive heat. This can be the result of abnormal combustion chamber temperatures or of a failure of the cooling system. Faulty injector timing, failure of the exhaust gas recirculation (EGR) or sustained high-speed/high-load operation system will create high cylinder temperatures. The thermal path from the valve face to the water jacket can be blocked by insufficient lash (which holds the valve off its relatively cool seat), local water-jacket corrosion, or a malfunction associated with flow directors. Also known as diverters, flow directors are inserted into the water jacket to channel coolant to the valve seats and injector sleeves. These parts can corrode or vibrate loose. Of course, a generalized failure of the cooling system, affecting all cylinders, is possible.

Valve breakage usually takes the form of fatigue failure from repeated shock loads or bending forces. Fatigue failure leaves a series of rings, not unlike tree growth rings, on the parting surfaces. Shock loads are caused by excessive valve lash, which slams the valve face against the seat, or weak valve springs. Bending forces are generated if the seat and/or guide is not concentric with the valve face.

The most dramatic form of valve failure comes about because of collision with the piston. As mentioned previously, Navistar allows as little as 0.009 in. between the exhaust valve and the piston crown at convergence, which occurs $4\frac{1}{4}°$ btdc. Admittedly, the 6.9 is a "tight" engine, but no diesel will tolerate excessive valve lash, weak valve springs, or overspeeding.

Valve service involves two operations, both of which are done by machine. The face is reground to the specified angle, which is usually (but not always) 30° on the intake and 45° on the exhaust, less a small angle, usually about 1.5°. Thus, the 30° intake seats specified for Bedford engines (Figure 7-29) would be cut to 28.5°. A valve face grinder adjusts to any angle. Seats are cut true to specification, which creates a slight mismatch, known as the interference angle. Once the engine starts, valve faces pound into conformity with the seats. The interference angle ensures a gastight seal on initial start-up and eliminates the need for lapping.

Seats are typically ground in three angles: an *entry angle*, which ranges from about 60°–70°, the factory-specified *seat angle*, and the *exit angle*, which is usually 15°. Thus seat width can be controlled by undercutting either of the flanking angles (Fig. 7-30). The seat should center on the valve face.

Intake

A. Valve seating angle: 30°
B. Valve seating width: 0.055–0.069 in.
C. Valve seat angle: 29°
D. Valve head minimum thickness: 0.035 in. valve head depth in relation to cylinder head face (minimum permissisble): 0.023 in.

A

Exhaust

E. Valve seating width: 0.097–0.110 in.
F. Valve seating angle: 45°
G. Valve head minimum thickness: 0.035 in.
Valve head depth in relation to cylinder head face (minimum permissible): 0.041

B

7-29 Typical valve specifications, Bedford engines.

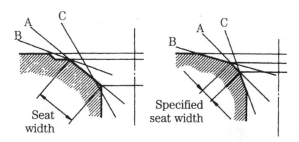

Before correction   After correction

7-30 Valve seat details—Yanmar GM series engines.

Depth gauge

7-31 A technician measures valve protrusion with a depth micrometer. Yanmar Diesel Engine Co., Ltd.

Valve protrusion or, as the case might be, recession, must be held to tight limits on these engines. A valve that extends too far into the chamber might be struck by the piston; one that is sunk into its seat hinders the combustion process, sometimes critically. Figure 7-31 illustrates how the measurement is made. Recession can be cured by replacing the valve and/or seat. Protrusion is a more difficult problem, encountered when the head has been resurfaced; a deep valve grind might help, provided one has the requisite seat thickness and can tolerate the increase in installed valve height (see immediately below). A handful of OEMs supply thinner-than-stock seat inserts. These inserts require corresponding thicker valve spring seats or shims, in order to maintain valve spring tension. Ultimately, one might be forced to replace the cylinder head.

Installed valve height, measured from the valve seat to the underside of the spring retainer, or cap, has become increasingly significant (Figure 7-32). Springs on newer engines come perilously close to coil bind with as little as 0.005 in. separating the coils in the full open position.

Finally, the machinist dresses the valve stem tip square to restore rocker-tip to increase available contact area and to center hydraulic-lifter pistons. The average lifter has a stroke of 0.150 in. If the plunger is centered, available travel is 0.075 in., before it bottoms and holds the valve off its seat. But it is rare to find new engines with precisely centered lifter plungers, and available stroke might be considerably less than indicated. Deep valve grinding, head and block resurfacing, and camshaft grinding will, unless compensated for, collapse the lifters. The machinist can remove about 0.030 in. from the tip without penetrating the case hardening. Another limiting factor is rocker/spring-cap contact.

Seat replacement

Valve seat inserts are pressed into recesses machined into the head. (Some very early engines used spigoted inserts with mixed results.) Seats must be replaced when severely burned, cracked, loose, or as a means of obtaining the correct valve protrusion. Many shops routinely replace seats during an overhaul for the insurance value.

7-32 Installed valve height is the distance between the spring seat and the underside of the spring retainer. Often the factory neglects to provide the specification, and it must be determined by direct measurement during disassembly.

Seats in iron heads are customarily driven out with a punch inserted through the ports, although more elegant tools are available (Figure 7-33); seats in aluminum heads should be cut out to prevent damage to the recess (Figure 7-34).

While new seats can be installed in the original counterbores, it is good practice to machine the bores to the next oversize. Replacement seats for most engines are available in 0.010-, 0.015-, 0.020-, and 0.030-in. oversizes. Material determines the fit: iron seats in iron heads require about 0.005-in. interference fit; Stellite expands less with heat, and seats made of this material should be set up a little tighter; seats in aluminum require something on the order of 0.008-in. interference.

Seat concentricity should be checked with a dial indicator mounted in a fixture that pilots on the valve guide. Often the technician finds it necessary to restore concentricity by lightly grinding the seat.

Springs

Valve spring tension is all that keeps the valves from hitting the pistons. A "swallowed" valve is the mechanic's equivalent of the great Lisbon earthquake or the gas blowout at King Christian Island, which illuminated the Arctic night for eight months and could be seen from the moon. Thus, I suggest that valve springs be replaced (regardless of apparent condition) during upper engine overhauls.

If springs are to be used, inspect as follows:

- Carefully examine the springs for pitting, flaking, and flattened ends.
- Measure spring freestanding height and compare with the factory wear limit.

7-33 Chrysler-Nissan supplies this valve seat puller. The split jaws (5) can open as the nut (4) is tightened. Tightening the upper nut (2) lifts the seat.

- Stand the springs on their ends and, using a feeler gauge and a machinist's square, determine the offset of the uppermost coil (Figure 7-35). Compare with the factory specification (in angular terms, maximum allowable tilt rarely exceeds 2°). Most keeper failures arise from unequal loading.
- Verify that spring tension falls within factory-recommended norms. Figure 7-36 illustrates the tool generally used to make this determination.

Valve spring shims have appropriate uses, chiefly to restore the spring preload lost when heads and valve seats are refurbished. But shims should not be used as a tonic for tired springs, because the fix is temporary and can result in coil bind.

7-34 Seats in aluminum heads should be machined out, rather than forced out. Onan

7-35 Most engine makers provide a linear valve spring tilt specification; Yanmar is more sophisticated. First spring free length is determined as shown in the left-hand drawing. The amount of offset is measured with a machinist's square. Offset (B) divided by free length (A) gives the specification, which for one engine series must not exceed 0.0035 in.

Top clearance

Top clearance, or the piston-to-head clearance at tdc, is critical. Unfortunately, the position of piston crown varies somewhat between cylinders because of the stacked tolerances at the crankshaft, rod, piston pin bosses, and deck (which might be tilted relative to the crankshaft centerline). No two pistons have the same spatial relationship to the upper deck. Resurfacing, or decking, the block lowers the fire deck 0.010 in. or more with no better accuracy than obtained by the factory.

7-36 Most valve spring testers measure force with a torque wrench. Ford Motor Co.

Most manufacturers arrive at top clearance indirectly by means of a piston deck height specification. Either of these measurements must be made when:

- the block is resurfaced.
- the manufacturer supplies replacement head gaskets in varying thickness to compensate for production variations.

The geometry of some engines (flat pistons and access to the piston top with head in place) invites direct measurement of top clearance. The cylinder head is installed and torqued, using a new gasket of indeterminate thickness. The mechanic then removes No. 1 cylinder glow plug and inserts the end of a piece of soft wire, known as fuse wire, into the chamber. Turning the flywheel through tdc flattens the wire between the piston crown and cylinder head. Wire thickness equals top clearance, or piston deck height plus compressed gasket thickness (Figure 7-37).

Scrupulous engine builders sometimes make the same determination using modeling clay as the medium. Upon disassembly the clay is removed and carefully miked. This method applies equally well to flat and domed pistons.

The more usual approach is to measure piston deck height with a dial indicator. The procedure involves three measurements, detailed as follows:

1. Zero the dial indicator on the fire deck, with the piston down (left-hand portion of Figure 7-38).
2. Position the indicator over a designated part of the piston crown, shown as A in Figure 7-38.

7-37 Top clearance equals piston deck height plus the thickness of the compressed head gasket. Yanmar Diesel Engine Co. Ltd.

7-38 The first step when measuring piston deck height is to zero the dial indicator on the fire deck. The second step in the measurement process is to determine piston protrusion at two points, the average of which gives piston protrusion. Ford Motor Co.

3. Turn the crankshaft in the normal direction of rotation through tdc. Note the highest indicator reading.
4. Repeat Step 3, taking the measurement at B.
5. Average measurements A and B.
6. Repeat the process for each piston. Use the piston with the highest average deck height to determine the thickness of the replacement head gasket.

Assembly

The manufacturer's manual provides detailed assembly instructions but includes little about the things that can go wrong. Most assembly errors can be categorized as follows:

- *Insufficient lubrication.* Heavily oil sliding and reciprocating parts, lightly oil head bolts and other fasteners, except those that penetrate into the water jacket. These fasteners should be sealed with Permatex No. 2 or the high-tech equivalent.

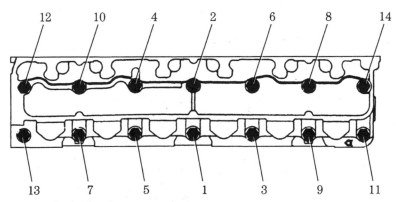

7-39 Cylinder head torque sequence. Gm Bedford Diesel

7-40 Valve lash can be measured at any accessible point in the system. For example, lash for the Ford-supplied 2.4L ohc is read as cam-to-rocker clearance (A); Onan measures the clearance between rocker arm tip and valve stem (B).

- *Reversed orientation.* Most head gaskets, many head bolt washers, and all thermostats are asymmetrical.
- *Mechanical damage.* Run fasteners down in approved torque sequences and in three steps—1/2, 2/3, and 1/1 torque (Figure 7-39). Exceptions are torque-to-yield head bolts and rocker arm shaft fasteners. The former are torqued as indicated by the manufacturer, whose instructions will be quite explicit. The latter—rocker shaft fasteners—should be brought down in very small increments, working from the center bolts out.

Gaskets, especially head gaskets, might also be damaged during assembly. Lower the head on a pair of guide pins lightly threaded into the block. Pins can be fashioned from discarded head bolts by cutting the heads off. If pins are too short to extend through the head casting, slot the ends for screwdriver purchase.

Set initial valve lash adjustments, bleed the fuel system, start the engine. Final lash adjustments are usually made hot, after the engine has run for 20 minutes or so on the initial settings (Figure 7-40).

8
CHAPTER

Engine mechanics

A mechanic needs to be a part-time electrician, semipro fuel system specialist, self-taught millwright, amateur machinist, and back-bench welder. But what he or she is supposed to do is to rebuild engines, the subject of this chapter.

Scope of work

Block assemblies can be *repaired, overhauled,* or *rebuilt* (Figure 8-1). Spot repairs are either triggered by local failure (e.g., a sticking oil pressure relief valve or a noisy valve train bearing) or by a need to extract a few more hours from a worn-out engine. Many a poor mechanic has replaced an oil pump more out of hope than conviction.

While an overhaul is also an exercise in parts replacement, the scope is wider and usually occasioned by moderate cylinder and crankshaft-bearing wear. At the minimum, an overhaul entails grinding the valves and replacing piston rings and bearing inserts and whatever gaskets have been disturbed. The effort might extend to a new oil pump, timing and accessory drive parts, oil seals, cylinder liners (when easily accessible), together with new piston, ring, and wrist pin sets. Because the block and crankshaft remain in place, machine work is necessarily limited to the cylinder head.

In the classic sense, rebuilding an engine means the restoration of every frictional surface to its original dimension, alignment, and finish. The engine should theoretically be as good as new, or even better than new in the sense that used castings tend to hold dimension better than "green" parts. (Repeated heating and cooling cycles relieve stresses introduced during the casting process.) In addition, an older engine might benefit from late-production parts.

Although some mechanics would disagree, the rebuilding process cannot repeal the law of entropy. A competently rebuilt engine will be durable over the long run and will be reasonably reliable in the short term, but it will not quite match the factory norms. Subsurface flaws will not be detected. Metal lost to water jacket corrosion is irretrievably lost. Nor can original deck height, timing gear mesh, main bearing cap height, and camshaft geometry be achieved in any commercially practical sense. And

1 Cam follower
2 Camshaft
3 Cylinder block
4 Crankshaft
5 Crankshaft
  counterweight
6 Oil pan
7 Connecting rod
8 Liner packing rings
9 Cylinder liner
10 Piston
11 Piston pin
12 Piston rings
13 Oil pump

8-1 Sectional view of Deere 6076 block assembly and nomenclature.

the potential for error, on the part of both the machinist and the assembler, affects reliability. More often than not, a freshly rebuilt engine will experience "teething" difficulties.

On the other hand, the cost should not exceed half of the replacement cost, and engine life will be nearly doubled.

Traditionally, the work is divided among operator mechanics, who remove the engine from, its mounts, dismantle it, and consign the components to a machinist for inspection and refurbishing. The machinist might supply some or all of the replacement parts, which, together with the reworked parts, are returned to the mechanics for final assembly.

This approach organizes the work around specialists' skills, keeps the critical business of assembly in-house (where it probably belongs), and minimizes out-of-pocket expenses. One working mechanic—not the shop foreman—should have undiluted responsibility for the job, a responsibility that includes new and refurbished parts quality control (QC), assembly, installation, and start-up.

Engine machine work is an art like gunsmithing or watch repair in the sense that proficiency comes slowly, through years of patient application. In my anachronistic opinion, the best work comes out of small shops, where Model T crankshafts stand in racks, waiting for customers that never come, and the coffee pot hasn't been cleaned since 1940. These shops, in short, are places where a mill means a Bridgeport, a grinder is a Landis, and the lathes were made in South Bend.

Diagnosis

Before you begin you should have a good idea—or at least a plausible theory—about the nature of the problem. The diagnostic techniques described in Chap. 4 indicate whether or not major work is in order and, when supplemented by oil analysis, will suggest which class of parts—rings, gears, soft metal bearings, and so on—are wearing rapidly.

Test/analysis data, combined with a detailed operating history, should fairly well pinpoint the failure site (cylinder bore, crankshaft bearings, accessory drive, and the like). But analysis should not stop with merely verbal formulations. For example, it is hardly meaningful to say that a bearing or a piston ring set has "worn out" or "overheated." One should try to identify what associated failure or special operating condition selected those parts to fail. City-bus wheel bearings are a good example of the selection process; experience shows that the right front bearings tend to fail more often than those on the left. Traces of red oxide in the lubricant suggest that failure comes about because of moisture contamination, (i.e., water splash), which is more likely to occur on the curb side of the vehicle.

Once the mechanic understands the failure mechanism, it might be possible to correct matters by performing more frequent maintenance, upgrading parts quality, or modifying operating conditions.

Rigging

Figure 8-2 illustrates the proper lifting tackle for a large engine. Note how the spreader bar and adjustable crossheads keep the chains vertical to hook loads.

The engine shown incorporates lift brackets; less serious engines do not, and the mechanic is left to his own devising. In general, attachment points should straddle the center of gravity of the engine in two planes, so that it lifts horizontally without tilting. Chains must clear vulnerable parts, such as rocker covers and fuel lines. Lift brackets made of ⅜ in. flame-cut steel plate are the ideal, but *forged* eyebolts (available from fastener supply houses) are often more practical. A spreader bar will eliminate most bending loads, but there are times when chain angles of less than 90° cannot be entirely avoided. Reduce the bending load seen by the eyebolt with a short length of pipe and heavy washer, as shown in Figure 8-3.

8-2 A proper engine sling is a necessity. Detroit Diesel

¹/₈ in. (3 mm) thick
flat washer

Pipe

8-3 Hardened eyebolts occasionally must be used in lieu of lift brackets. Bolts should thread into a minimum depth of three times the diameter and should be reinforced as shown to limit bending forces.

8-4 Purchase the best engine stand you can afford, with a weight capacity that provides a comfortable safety margin.

In no case should chains be bolted directly to the block without the intermediary of a bracket or eyebolt. Nor is multistrand steel cable appropriate for this kind of knock-about service.

Figure 8-4 illustrates a minimal engine stand, suitable for engines in the 600-lb. class. Better stands usually attach at the side of the block (as opposed to the flywheel flange) and can be raised and lowered.

Note: A mechanic can get into trouble with one of these revolving-head stands. Inverting an assembled engine with the turbocharger intact can dump oil from the turbocharger sump into one or more cylinders. Subsequent attempts to start the flooded engine might result in bent connecting rods or worse.

Special considerations

Mechanics who are knowledgeable about gasoline engine repair will find themselves doing familiar things, but to more demanding standards. Diesel engines are characterized by:

Close tolerances Close tolerances impose severe requirements in terms of inspection, cleanliness, and torque limits. Tolerance stack—unacceptable variations in dimension of assemblies made up of components that fall on the high or low ends of the tolerance range—becomes a factor to contend with. As delivered from the factory, some engines employ a selective assembly of bearing inserts and pistons.

High levels of stress Hard-used industrial engines are subject to structural failures, a fact that underscores the need for careful inspection of crankshafts, main-bearing webs and caps, connecting rods, pistons, cylinder bore flanges, harmonic balancers, and all critical fasteners.

Inflexible refinishing and assembly norms Stress and close tolerances give little latitude for quick fixes and, unless contradicted by hard experience, factory recommendations should be followed to the letter.

Special features While nothing in four-cycle diesel crankcases can be considered uniquely diesel, some features of these engines might be unfamiliar to most gasoline-engine mechanics. Nearly all engines employ oil-cooled pistons. This is accomplished with rifle-drilled connecting rods or, as is more often the case today, by means of oil-spray tubes, or jets, which direct a stream of oil to the piston undersides (Figure 8-5). Jets that bolt or press into place must be removed for cleaning and usually require alignment.

Diesel engines often employ removable cylinder sleeves, pressed or slipped into counterbores and standing proud of the fire deck. Installation is quite critical and special honing techniques are usually recommended.

Another difference is the apparent complexity of power transmission, which on some engines can have an almost baroque complexity. Figure 8-6 illustrates the bull gear/idler gear and accessory drive gear constellation on the DDA Series 60 engine. Belt drives may be hardly less imposing, as the drawing in Figure 8-7 indicates.

In fact, this complexity is more illusory than real. One merely comes to terms with timing marks, deals with one component at a time, and builds the power

Piston cooling jet

8-5 Spray jet aiming is critical.
Ford Motor Co.

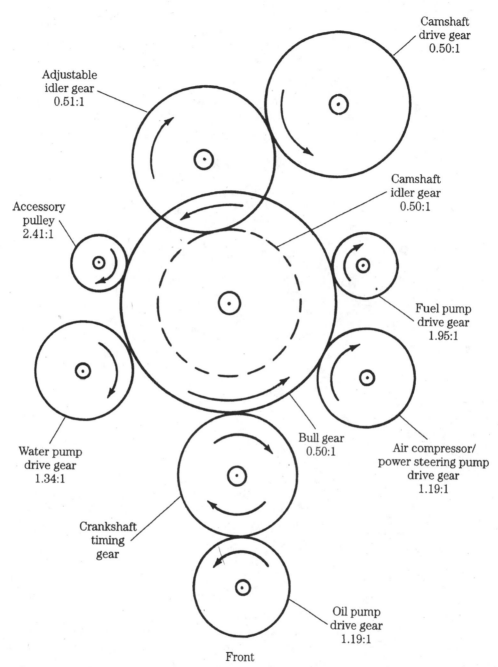

Camshaft
drive gear
0.50:1

Adjustable
idler gear
0.51:1

Camshaft
idler gear
0.50:1

Accessory
pulley
2.41:1

Fuel pump
drive gear
1.95:1

Water pump
drive gear
1.34:1

Bull gear
0.50:1

Air compressor/
power steering pump
drive gear
1.19:1

Crankshaft
timing
gear

Oil pump
drive gear
1.19:1

Front

8-6 Detroit Diesel engines are known for their sophisticated gear trains.

8-7 Ford 2.3L turbocharged diesel makes extensive use of toothed belts.

train brick by brick. Of course, the work goes more slowly than it would on a simpler engine, and the parts cost can be daunting, especially when gears need to be replaced.

The power train can include one or more balance shafts, a technology that becomes common on spark ignition engines (the Mitsubishi/Chrysler 2.6L is an example).

Figure 8-8 illustrates the balance shaft configuration for a four-cylinder in-line engine. Two contra-rotating shafts, labeled *Silent shafts* in the drawing, run at twice crankshaft speed to generate forces that counter the "natural" vibration of the engine.

Engines of this type employ single-plane, two-throw crankshafts. Pistons 1 and 4 move in concert, as do pistons 2 and 3. When pistons 1 and 4 are down, 2 and 3 are up. Consequently, vertical forces generated by pistons 1 and 4 (indicated by the dotted line in Figure 8-9) oppose the forces generated by the center pair of pistons (represented by the fine line). These vertical forces, known as primary shaking forces, almost cancel and can be ignored.

Secondary forces pose a more serious problem. Created by connecting-rod angularity and by piston acceleration during the expansion stroke, these forces tend both to rotate the engine around the crankshaft centerline and to shake the engine vertically. Magnitude increases geometrically with rpm, to produce the dull rattle characteristic of in-line four-cylinder engines at speed.

8-8 Balancing secondary forces requires two counterweighted shafts running at twice crankshaft speed. Ford Motor Co.

Secondary vertical and rolling forces are neutralized by deliberately induced imbalances in the balance shafts. Figure 8-10 diagrams the sequence of countervailing forces through full crankshaft revolution (two balance shaft revolutions).

Detroit Diesel approaches the question of balance differently on its two-cycle engines. Here the concern is to balance the rocking couple created by crankpin offset. Such couples exist unless all pistons share the same crankpin, as for example, in a radial engine with one connecting rod articulated from a central master rod. DDA practice is to use a counterweight can and balance shafts driven at crankshaft speed through counterweighted gears. In other words, each shaft has two sources of imbalance, one integral with the drive gear and the other in the form of a bob weight on

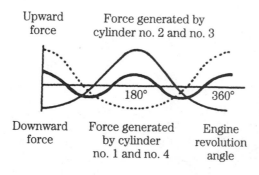

8-9 Primary forces in in-line four-cylinder engine very nearly balance. In an opposed four, with two crankshaft throws in the same plane 180° apart, balance is nearly perfect. A slight rocking couple, imposed by the offset between paired connecting rods on each crankshaft throw, does, however, exist. Ford Motor co.

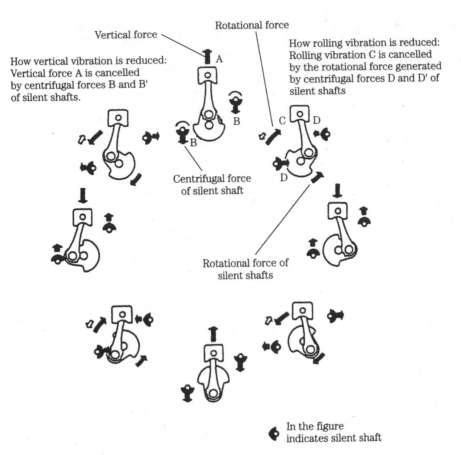

Vertical force

Rotational force

How vertical vibration is reduced:
Vertical force A is cancelled
by centrifugal forces B and B'
of silent shafts.

How rolling vibration is reduced:
Rolling vibration C is cancelled
by the rotational force generated
by centrifugal forces D and D' of
silent shafts

Centrifugal force
of silent shaft

Rotational force of
silent shafts

In the figure
indicates silent shaft

8-10 Balance shaft/crankshaft forces during a complete engine revolution. Ford Motor co.

the free end of the shaft. Shaft counterweights and shaft gear weights are disposed radially to create a countervailing couple, which acts in opposition to the crankshaft-induced couple. No attempt is made to balance secondary forces.

From a mechanic's point of view, the critical aspects of this technology are the shaft bearings, which are subject to severe radial loads and catastrophic failure. In some cases, bearing bosses must be sleeved before new bearings can be installed. Give the oiling circuit close scrutiny; endemic bearing failure can justify modifications to increase the rate of oil flow. And of course it is necessary to time the shafts relative to each other and to the crankshaft.

Fasteners

Contemporary foreign and, to a great extent, American engines are built to the metric ISO (International Standards Organization) standards, developed from the European DIN. For most practical purposes DIN and ISO fasteners interchange. A JIS standard also exists, but most Japanese fasteners made since the early 1970s follow the ISO pattern. Some JIS bolts interchange (although head dimension can differ) with those built to the current standard; others make up just enough to strip out.

Few American manufacturers remain wedded to the inch standard, although leaving is hard to do. Engines come off the line with both ISO and fractional fasteners, inch-standard pipe fittings and metric fuel systems.

ISO fasteners are classed by nominal bolt diameter in millimeters and thread pitch measured as millimeters between adjacent thread crests. Thus, a specification might call for an M8 × 1.0 cap screw or stud. Wrench size markings reflect bolt diameter, not the flat-to-flat distance across the screw head. Yield strength is indicated by a numeric code embossed on the screw head. The higher the number, the stronger the cap screw. Metric hex nuts often carry the same numerical code, but this practice is not universal.

Figures 8-11 and 8-12 supply identification data and suggested torque limits for U.S. and metric cap screws. Torque limits were calculated from bolt yield strength ratings, and do take into account the effects of clamping forces on vulnerable parts or gaskets. Consequently, these values should not be used when the engine manufacturer provides a different torque limit or tightening procedure for a specific application. Tighten plastic insert- or crimped steel-type locknuts to about half the amount shown in the charts; toothed- or serrated-type locknuts receive full torque. Replace fasteners with the same or higher grade, except in the case of shear bolts, which are grade specific. When substituting a better-grade fastener, torque it to the value of the original.

SAE Grade	Head Markings	SAE Grade	Nut Markings
SAE GRADE 1 / SAE GRADE 2	No Mark	2	No Mark

SAE Grade	Head Markings	SAE Grade
SAE GRADE 5 / SAE GRADE 5.1 / SAE GRADE 5.2	(markings)	5 Nut Markings

SAE Grade	Nut Markings	SAE Grade
SAE GRADE 8 / SAE GRADE 8.2	(markings)	8 Nut Markings

DIA.	WRENCH SIZE	SAE GRADE 1		*SAE GRADE 2		SAE GRADE 5		SAE GRADE 8	
		OIL	DRY	OIL	DRY	OIL	DRY	OIL	DRY
		N•m(lb-in)	N•m(lb-in)	N•m(lb-in)	N•m(lb-in)	N•m(lb-in)	N•m(lb-in)	N•m(lb-in)	N•m(lb-in)
#6 #8		0.5(4.5) 0.9(8)	0.7(6) 1.2(11)			1.4(12) 2.4(21)	1.7(15) 3.2(28)		
#10 #12		1.4(12) 2(19)	1.8(16) 2.8(25)			3.4(30) 5.4(48)	4.6(41) 7.3(65)		
		N•m(lb-ft)	N•m(lb-ft)	N•m(lb-ft)	N•m(lb-ft)	N•m(lb-ft)	N•m(lb-ft)	N•m(lb-ft)	N•m(lb-ft)
1/4 5/16	7/16 1/2	3.5(2.5) 7(5)	4(3) 9(6.5)	5(4) 10(7.5)	7(5) 14(10)	8(6) 16(12)	11(8) 23(17)	12(8.5) 24(18)	16(12) 33(24)
3/8 7/16	9/16 5/8	12(8.5) 19(14)	16(12) 26(19)	19(14) 30(22)	24(18) 41(30)	30(22) 47(35)	41(30) 68(50)	41(30) 68(50)	54(40) 95(70)
1/2 9/16	3/4 13/16	24(21) 41(30)	41(30) 54(40)	47(35) 68(50)	61(45) 88(65)	75(55) 108(80)	102(75) 142(105)	102(75) 149(110)	142(105) 203(150)
5/8 3/4	15/16 1-1/8	54(40) 102(75)	75(55) 136(100)	88(65) 163(120)	122(90) 217(160)	149(110) 258(190)	197(145) 353(260)	203(150) 366(270)	278(205) 495(365)
7/8 1	1-5/16 1-1/2	163(120) 244(180)	244(165) 332(245)	163(120) 244(180)	224(165) 332(245)	414(305) 624(460)	563(415) 848(625)	590(435) 881(650)	800(590) 1193(880)
1-1/8 1-1/4	1-11/16 1-7/8	346(255) 488(360)	468(345) 664(490)	346(255) 488(360)	468(345) 665(490)	780(575) 1098(810)	1058(780) 1492(1100)	1248(920) 1763(1300)	1695(1250) 2393(1765)
1-3/8 1-1/2	2-1/16 2-1/4	637(470) 848(625)	868(640) 1153(850)	637(470) 848(625)	868(640) 1153(850)	1438(1061) 1912(1410)	1953(1440) 2590(1910)	2312(1705) 3065(2260)	3140(2315) 4163(3070)

8-11 Inch cap screw torque values.

Property Class and Head Markings	4.6		4.8		8.8	9.8	10.9		12.9	
Property Class and Nut Markings	5		5		10		10		12	

DIA.	WRENCH SIZE	4.6		4.8		8.8 or 9.8		10.9		12.9	
		OIL	DRY	OIL	DRY	OIL	DRY	OIL	DRY	OIL	DRY
		N•m(lb-ft)	N•m(lb-ft)	N•m(lb-ft)	N•m(lb-ft)	N•m(lb-ft)	N•m(lb-ft)	N•m(lb-ft)	N•m(lb-ft)	N•m(lb-ft)	N•m(lb-ft)
M3	5.5mm	0.4(0.2)	0.5(0.3)	0.5(0.4)	0.7(0.5)	1(0.8)	1.3(1)	1.5(1)	2(1.5)	1.5(1)	2(1.5)
M4	7mm	0.9(0.6)	1.1(0.8)	1(0.9)	1.5(1)	2.5(1.5)	3(2)	3.5(2.5)	4.5(3)	4(3)	5(4)
M5	8mm	1.5(1)	2.5(1.5)	2.5(1.5)	3(2)	4.5(3.5)	6(4.5)	6.5(4.5)	9(6.5)	7.5(5.5)	10(7.5)
M6	10mm	3(2)	4(3)	4(3)	5.5(4)	7.5(5.5)	10(7.5)	11(8)	15(11)	13(9.5)	18(13)
M8	13mm	7(5)	9.5(7)	10(7.5)	13(10)	18(13)	25(18)	25(18)	35(26)	30(22)	45(33)
M10	16mm	14(10)	19(14)	20(15)	25(18)	35(26)	50(37)	55(41)	75(55)	65(48)	85(63)
M12	18mm	25(18)	35(26)	35(26)	45(33)	65(48)	85(63)	95(70)	130(97)	110(81)	150(111)
M14	21mm	40(30)	50(37)	55(41)	75(55)	100(74)	140(103)	150(111)	205(151)	175(129)	240(177)
M16	24mm	60(44)	80(59)	85(63)	115(85)	160(118)	215(159)	235(173)	315(232)	275(203)	370(273)
M18	27mm	80(59)	110(81)	115(85)	160(118)	225(166)	305(225)	320(236)	435(321)	375(277)	510(376)
M20	30mm	115(85)	160(118)	165(122)	225(166)	320(236)	435(321)	455(356)	620(457)	535(395)	725(535)
M22	33mm	160(118)	215(159)	225(167)	305(225)	435(321)	590(435)	620(457)	840(620)	725(535)	985(726)
M24	36mm	200(148)	275(203)	285(210)	390(288)	555(409)	750(553)	790(583)	1070(789)	925(682)	1255(926)
M27	41mm	295(218)	400(295)	415(306)	565(417)	810(597)	1100(811)	1155(852)	1565(1154)	1350(996)	1835(1353)
M30	46mm	400(295)	545(402)	565(417)	770(568)	1100(811)	1495(1103)	1570(1158)	2130(1571)	1835(1353)	2490(1837)
M33	51mm	545(402)	740(546)	770(568)	1050(774)	1500(1106)	2035(1500)	2135(1575)	2900(2139)	2500(1844)	3390(2500)
M36	55mm	700(516)	950(700)	990(730)	1345(992)	1925(1420)	2610(1925)	2740(2021)	3720(2744)	3205(2364)	4355(3212)

8-12 Metric cap screw torque values. 1990.

Cleaning

Ford Motor and other manufacturers say that dirt is the chief cause of callbacks after major repair work. The direct effect is to contaminate the oil supply; the indirect effect is to create an environment that makes craftsmanship difficult or impossible.

The need for almost septic standards of cleanliness argues against the practice of opening the engine for less than comprehensive repairs. Of course, it happens that such repairs must be made, regardless of the long-term consequences. Nor is it possible to maintain reasonable standards of cleanliness during in-frame overhauls, although the damage can be minimized by moving in quickly, cleaning only those friction surfaces that are opened for inspection, and getting out. Dirt accumulations on internal parts of the engine cannot be removed from below, while parts are still assembled, and attempts to do so will only release more solids into the oil stream.

When, on the other hand, the engine is rebuilt, the block, cylinder head, pan, and other steel stampings are sent out for thermal or chemical cleaning (see Chap. 7). These processes also remove the paint, which is all to the good. Crankshaft, piston assemblies, and other major internal parts receive a preliminary wash-down for inspection by the mechanic in charge of the job. These parts are then forwarded to the machine shop for evaluation. When pistons are reused, the machinist will chemically clean the grooves and piston undersides—chores that save hours of labor. About all that remains for the shop mechanic is to degrease fasteners and accessories that have been detached from the block and remove the rust preventative from new parts.

Teardown

Drain the oil and coolant and degrease the outer surfaces of the engine. Disconnect the battery, wiring harness (make a sketch of the connections if the harness is not keyed for proper assembly), and exhaust system. Attach the sling and undo the drive line connection and the motor mounts. With the engine secured in a stand, detach the manifolds, cylinder heads, and oil sump. The block should be stripped if you contemplate machine work or chemical cleaning of the jackets.

Lubrication system

The first order of business should be the lubrication system. To check it out you must have a reasonably good notion of the oiling circuits. Figure 8-13 is a

8-13 Lubrication system. Onan

drawing of the Onan DJ series lubrication system. The crankcase breather is included because it has much to do with oil control. Should it clog, the engine will leak at every pore. Oil passes from the screened pickup tube (suspended in the pan) to the pump, which sends it through the filter. From there the filtered oil is distributed to the camshaft, the main bearings (and through rifle-drilled holes in the crankshaft to the rod bearings and wrist pin), and to the valve gear. Valve and rocker arm lubrication is done through a typically Onan "showerhead" tube. Tiny holes are drilled in the line and deliver an oil spray at about 25 psi. On its return to the sump, the oil dribbles down the push rods to lubricate the cam lobes and tappets.

The oiling system in Figure 8-14 is typical of larger engines. From the bottom of the drawing, oil enters the pickup tube, then goes to the Gerotor pump. Unlike most oil pumps, this one is mounted on the end of the crankshaft and turns at engine speed. The front engine cover incorporates inlet and pump discharge ports. Oil is sent through a remote cooler (6), then directed back to the block, where it exits again to the filter bank (11). Normally oil passes through these filters. However, if the filters clog or if the oil is thick from cold, a pressure differential type of bypass valve (12) opens and allows unfiltered flow.

The main oil gallery (3) distributes the flow throughout the engine. Some goes to the main bearings and through the drilled crankshaft to the rods. The camshaft

8-14 A more complex lubrication system. International Hamster

bearings receive oil from the same passages that feed the mains. The rearmost cam journal is grooved. Oil passes along this groove and up to the rocker arms through the hollow rocker shaft. On its return this oil lubricates the valve stems, pushrod ball sockets, tappets, and cam lobes. Other makes employ similar circuits, often with a geared pump.

Figure 8-15 describes the Ford 2.3L lubrication system, which is surprisingly sophisticated for a small and, by diesel standards, inexpensive product. Pressurized oil goes first to the filter, described in the following section.

Depending on oil temperature, output from the filter goes either directly to the main distribution gallery or to the gallery by way of an oil cooler (Figure 8-16). A thermostatic valve, located in the filter housing, directs the flow (Figure 8-17). Cooling the oil during cold starts is counterproductive, and the valve remains closed; as oil temperature increases, the valve extends to split the flow between the gallery and cooler. At approximately 94°C, all flow is diverted to the cooler. As a safety measure, the bypass valve opens if cooler pressure drop exceeds 14 psi. Thus, a stoppage will not shut down oil flow, although long-term survivability is compromised.

Camshaft lubrication, a weak point on many ohc engines, shows evidence of careful attention. A large-diameter riser, feeding from the main gallery, supplies oil to the center camshaft bearing. At that point the flow splits, part of it entering the hollow camshaft and part of it going to an overhead spray bar. The camshaft acts as an oil gallery for the remaining two shaft bearings; the spray bar provides lubrication for camshaft lobes, rockers, and valve tips.

The main bearings are lubricated through rifle-drilled ports connecting the webs and main gallery. As a common practice, diagonal ports drilled in the crankshaft convey oil from the webs to adjacent crankpins. Spray jets, again fed by the main gallery, cool the undersides of the pistons and provide oil for the piston pins. A gear and crescent pump—the lightest, most compact type available—supplies the necessary pressure.

This cursory examination of the oil circuitry on three very different engines should underscore the need to come to terms with these systems. No other engine system has such immediate implications for the integrity of the mechanic's work.

Clogging is the most frequent complaint, caused by failure to change oil and filters at proscribed intervals, and abetted by design flaws. Expect to find total or partial blockage wherever the oil flow abruptly changes direction or loses velocity. Prime candidates are the cylinder head/block interface, spray jets, and the junctions between cross-drilled passages. Neither immersion-type chemicals, heat, nor compressed air can be depended on to remove metal chips and sludge. Drilled passages must be cleared by hand, using riflebore brushes and solvent.

Thorough cleaning is never more important than after catastrophic failure, which releases a flood of metallic debris into the oiling system. In this case, the oil cooler can present a special problem. Modern, high-density coolers do not respond well to chemical cleaners and frustrate even the most flexible swabs. A new or used part—whose history is known—is sometimes the only solution.

Oil leaks can develop at the welch (expansion) plugs or, less often, at the pipe plugs that blank off cross-drilled holes and core cavities. It is also possible for cracks to open around lifter bores and other thin-section areas. A careful mechanic

A

8-15 Ford 2.4L lubrication system.

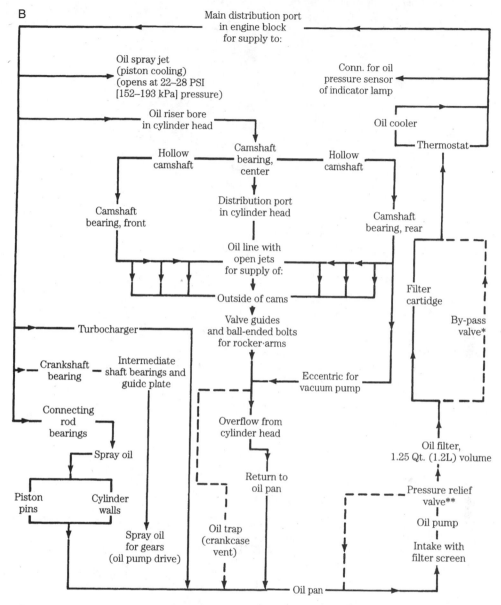

B

Main distribution port
in engine block
for supply to:

Oil spray jet
(piston cooling)
(opens at 22–28 PSI
[152–193 kPa] pressure)

Conn. for oil
pressure sensor
of indicator lamp

Oil riser bore
in cylinder head

Oil cooler

Thermostat

Hollow
camshaft

Camshaft
bearing,
center

Hollow
camshaft

Camshaft
bearing, front

Distribution port
in cylinder head

Camshaft
bearing, rear

Oil line with
open jets
for supply of:

Filter
cartridge

By-pass
valve*

Outside of cams

Valve guides
and ball-ended bolts
for rocker·arms

Turbocharger

Eccentric for
vacuum pump

Crankshaft
bearing

Intermediate
shaft bearings and
guide plate

Connecting
rod
bearings

Oil filter,
1.25 Qt. (1.2L) volume

Spray oil

Overflow from
cylinder head

Pressure relief
valve**

Piston
pins

Cylinder
walls

Return to
oil pan

Oil pump

Spray oil
for gears
(oil pump drive)

Oil trap
(crankcase
vent)

Intake with
filter screen

Oil pan

*(Opens at 36 PSI [241 kPa]) oil supply guaranteed if filter cartridge is plugged

**(Oil pump) with cold oil (opens with 79–92 PSI [544–648 kPa] pressure)

8-15 (*Continued*)

8-16 Oil cooler.

will discard welch plugs (after determining that spares are available!) and apply fresh sealant to pipe plug threads during a rebuild. Most oil-wetted cracks can be detected visually.

Internal leaks that have been missed will show when the system is filled with pressurized oil before start-up.

8-17 Thermostatic bypass valve.

Filters

All contemporary engines employ full-flow filtration, using single or tandem paper-element filters in series with the pump outlet. Pleated paper filters can trap particles as small as 1 μm but are handicapped by limited holding capacity. Consequently, such filters incorporate a bypass valve that opens to shunt the element when the pressure drop reaches about 15 psi (Figure 8-18), an action that should trigger a warning lamp on the instrument panel (Figure 8-19). Otherwise the operator should occasionally feel the filter canister to verify that warm oil is circulating.

The Ford 2.3L filter consists of two concentric elements, one in series with the oiling circuit, the other in parallel (Figure 8-20). As long as the bypass valve, shown at the top of the drawing, remains seated, pump output passes through the outer, series-connected element and into the lubrication circuit. Pump delivery rates are greater than the system requires and the surplus migrates through the inner filter and returns to the sump through the 9-mm orifice at the base of the assembly. I have been unable to obtain precise data for this engine, which was designed in Japan. However, standard design practice is to calculate diesel lube oil requirements as 0.003 times the ratio of oil to piston displacement times rpm. Pumps are then sized to meet twice these requirements. If this holds, the parallel filter makes a major contribution to oil cleanliness.

The most important single maintenance activity is to change the oil and filter on the engine-maker's schedule, which is more demanding than suggested for equivalent spark ignition (SI) engines. A turbocharger increases bearing loads and blowby, further shortening oil and filter life.

The final steps in an overhaul or rebuild are to prime the oil pump by rotating the shaft with the impeller submerged in a container of clean motor oil, fill the turbocharger oil cavity and feed line, and pressurize the oiling circuit. The latter is accomplished by connecting a source of oil pressure to the main distribution cavity, usually by way of the oil-pressure sender tap. This step is sometimes omitted and

8-18 Yanmar cartridge-type filter mounts downstream of the pressure-regulating valve and includes a bypass valve.

Sender

8-19 Peugeot automotive diesels alerted the driver when the filter clogged.

Bypass
97±21 kPa

Full flow
element

Bypass
element

9 mm
orifice

To
oil pan

To oil
gallery

From
oil pump

8-20 Ford series/parallel filter.

the bearings—even though preoiled during assembly—suffer for it. The system *must* be pressurized after a new camshaft is installed.

Any repairs involving the turbocharger should be followed by disconnecting the oil return line and cranking until flow is established.

Before start-up after routine oil and filter changes, it is good practice to dry-crank the engine until the panel gauge indicates that oil pressure has been restored.

Lube oil

Oil collects moisture from condensation and is the repository for the liquid by-products of combustion. These by-products include several acid families that, even in dilute form, attack bearings and friction surfaces. No commercially practical filter can take out these contaminants. In addition oil, in a sense, wears out. The petroleum base does not change, but the additives become exhausted and no longer suppress foam, retard rust, and keep particles suspended. Heavy sludge in the filter is a sure indication that the change interval should be shortened, because the detergents in the oil have been exhausted. In heavy concentrations water emulsifies to produce a white, mayonnaise-like gel that has almost no lubrication qualities.

Oil that has overheated oxidizes and turns black. (This change should not be confused with the normal discoloration of detergent oil.) If fresh oil is added, a reaction might be set up that causes the formation of hard granular particles in the sump, known as "coffee grounds."

Oil change intervals are a matter of specification—usually at every 100 hours—and sooner if the oil shows evidence of deterioration. Most manufacturers are quite specific about the type and brand to be used. Multigrade oils (e.g., 10–30W) are not recommended for some engines, because it is believed they do not offer the protection of single-weight types. Other manufacturers specify SF or higher grades. As a practical matter this specification means that multigrade oils can be used. Brand names are important in diesel service. Compliance with the standards jointly developed by the American Petroleum Institute, the Society of Automotive Engineers, and the American Society for Testing and Materials is voluntary. You have no guarantee that brand X is the equivalent of brand Y, even though the oil might be labeled as meeting the same API-SAE-ASTM standards. General Motors suggests that you discuss your lubrication needs with your supplier and use an oil that has been successful in diesel engines and meets the pertinent military standards. MIL-L-2104B is the standard most often quoted, although additives cause problems with some engines often in the form of deposits around the ring belt. For GM two-cycles zinc should be held in between 0.07% and 0.10% by weight, and sulfated ash to 1.0%. If the lubricant contains only barium detergent-dispersants, the sulfated-ash content can be increased by 0.05%. High-sulfur fuels might call for *low-ash series 3* oils, which do not necessarily meet military low-temperature performance standards. Such oils might not function as well as MIL-L-2104B lubricants in winter operation.

Oil pumps

The Gerotor pump shown in Figure 8-21 is gear-driven; a more common practice is to drive pumps of this type directly from the crankshaft. Like the Wankel engine, which employs a similar trochoidal geometry, operation is somewhat difficult to visualize on paper and obvious in the hardware. The inner rotor, which has one lobe less than the outer rotor, "walks" as it turns, imparting motion to the outer rotor and simultaneously varying the volume of the working cavity (shaded in the drawing). Volume change translates as a pressure head.

The clearance between the rotors and the end cover is best checked with *Plasti-Gage*. Plasti-Gage consists of rectangular-section plastic wire in various thicknesses. A length of the wire is inserted between the rotors and the pump cover, and the cover torqued down to specification. The working clearance is a function of how much the wire is compressed under assembly torque. The package has a scale printed on it to convert this width to thousandths of an inch. Plasti-Gage is extremely accurate and fast. The only precautions in its use are that the parts must be dry, with no oil or solvent adhering to them, and that the wire must be removed after measurement. Otherwise it might break free and circulate with the oil, where it can lodge in a port. Typical end clearance for a Gerotor pump is on the order of 0.003 in. Of course, all rubbing surfaces should be inspected for deep scratches, and the screen should be soaked in solvent to open the pores.

8-21 Gerotor pump. Yanmar Diesel Engine Co. Ltd.

Other critical specifications are the clearance between the outer rotor and the case (Figure 8-22A) and the approach distance between inner and outer rotors (B).

Camshaft-driven gear-type pumps are the norm (Figure 8-23). Check the pump for obvious damage—scores, chipped teeth, noisy operation. Then measure the clearance; the closer the better as long as the parts are not in physical contact.

End clearance can be checked by bolting the cover plate up with Plasti-Gage between the cover and gears, or with the aid of a machinist's straightedge. Lay the straightedge on the gear case and measure the clearance between the top of

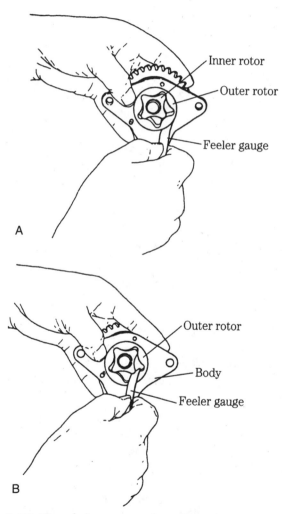

8-22 Use a feeler gauge to determine outer rotor to case clearance (A) and inner to outer rotor clearance. Wear limits for the Yanmar and most other such pumps are 0.15 in. (B).

1 Oil pump body
2 Idler shaft
3 Packing (SD22)
4 Oil pump cover
5 Drive gear
6 Drive shaft
7 Pin
8 Driven gear
9 Relief valve
10 Relief valve spring
11 Washer
12 Cotter pin
13 Oil screen
14 Bolt
15 Bolt
16 Washer
17 Lockwasher
18 Bolt
19 Gasket
20 Bolt
21 Lockwasher
22 Driving spindle
23 Driving-spindle support
24 O-ring
25 Bolt
26 Bolt
27 Lockwasher

8-23 Exploded view of typical gear-type oil pump. Marine Engine Div., Chrysler Corp.

the gear and the case with a feeler gage. When in doubt, replace or resurface the cover. Check the diameter of all shafts and replace as needed. The idler gear shaft is typically pressed into the pump body. Use an arbor press to install, and make certain that the shaft is precisely centered in its boss. Allowing the shaft to cant will cause interference and early failure. Total wear between the drive shaft and bushing can be determined with a dial indicator as shown in Figure 8-24. Move the shaft up and down as you turn it. Check the backlash with a piece of solder or Plasti-Gage between the gears. For most pumps this clearance should not exceed 0.018 in.

In general, it is wise not to tamper with the pressure relief valve, which might be integral with the pump located at some distance from it. If you must open the valve, because of either excessive lube oil pressure or low pressure, observe that the parts are generally under spring tension and can "explode" with considerable force. Check the spring tension and free length against specs, and replace or lap

B

8-23 (*Continued*)

8-24 Determining shaft-bearing wear. Marine Engine Div., Chrysler Corp.

8-25 Oil pressure check. Peugeot

the valve parts as needed. When this assembly has been disturbed it is imperative that the oil pressure be checked, with an accurate instrument such as the gauge shown in Figure 8-25.

Oil pressure monitoring devices

If the engine manufacturer has not done so, it makes sense to invest in a low oil pressure alarm to supplement the gauge or idiot light. Mechanical gauges fail or respond erratically if the gauge fitting is blocked by a slug of impacted oil. Electronic gauge or warning lamp failure is usually the fault of the oil-pressure sensor or related wiring.

Stewart-Warner monitor Stewart-Warner makes an audiovisual monitor that lights and buzzes when oil pressure and coolant temperatures exceed safe norms. The indicator is a cold-cathode electron tube, which is much more reliable than the usual incandescent bulb. Designed for panel mounting, the 366-T3 is relatively inexpensive insurance.

Onan monitor Onan engines can be supplied with an automatic shutdown feature. The system employs a normally ON sensor element and a time-delay relay. In conjunction with a 10 W, 1-Ω resistor, this relay allows the engine to be started. When running pressure drops below 13 psi, the sensor opens, causing the fuel rack to pull out.

8-26 Onan oil pressure monitor—sensor section.

The sensor (Figure 8-26) is the most complicated part of the system. It has a set of points that should be inspected periodically, cleaned, and gapped to 0.040 in. When necessary, disassemble the unit to check for wear in the spacer, fiber plunger, and spring-loaded shaft plunger. The spacer must be at least 0.35 in. long. Replace as needed. Check the action of the centrifugal mechanism by moving the weights in their orbits. Binding or other evidence of wear dictates that the weights and cam assembly be replaced. The cam must not be loose on the gear shaft.

Ulanet monitor The George Ulanet Company makes one of the most complete monitoring systems on the market (Figure 8-27). It incorporates an optional water level sensor, overheat sensor, fuel-pressure sensor, and low- and high-speed oil sensors.

Crankcase ventilation

The crankcase must he vented to reduce the concentration of acids and water in the oil. A few examples are at atmospheric pressure; others (employing forced-air scavenging) are maintained slightly higher than atmospheric pressure; and still other systems run the crankcase at a slight vacuum to reduce the possibility of air leaks. The vapors are recycled through the intake ports. In any event, the system requires attention to ensure that it operates properly. A clogged mesh element or vent pipe will cause unhealthy increases in crankcase pressures, forcing oil out around the gaskets and possibly past the seals, as well as increasing oil consumption. Figure 8-28

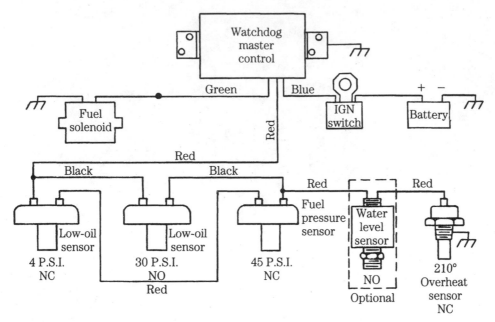

8-27 Ulanet monitor as configured for DD two-cycle engines.

shows a breather assembly with check valve to ensure that the case remains at less than atmospheric pressure.

Block casting

Carefully examine the fire deck for pulled head bolt threads, eroded coolant passages, missing or damaged coolant deflectors, and cracks, particularly between adjacent cylinders. In some cases, minor cracks can be ground out and filled. Several dimensional checks also need to be made at this time.

Deck flatness

Figure 8-29 illustrates the basic technique, which involves a series of diagonal and transverse measurements along the length and width of the deck, using a machinist's straightedge and feeler gauge. Block deck surfaces can he milled flat (a process called *decking*), although contemporary design practices give little leeway for corrective machining. As mentioned in Chap. 7, several modern engines cannot be safely decked, and warped blocks must be replaced. However, when it comes to it, most machinists will push these limits and obtain piston-to-cylinder head clearance by shimming the gasket or by selective assembly, fitting the shortest piston and rod sets that can be found. The owner should be aware of these expedients and agree to them.

8-28 Yanmar atmospheric crankcase breather.

Piston height

While this is an assembly dimension, an early check is not out of order. Measure piston protrusion or regression with a dial indicator, as described in the previous chapter.

8-29 If the head gasket is to live, the fire deck must be flat and true. Chrysler Corp.

Liner height

A few engines continue to be built with integral cylinder bores, machined directly into the block metal or in the form of a sleeve permanently installed during the casting process. These engines have flat decks, and what will be said here does not apply.

Most diesel engines employ discrete cylinder bore liners, or sleeves, which can be more or less easily replaced. Dry liners press into block counterbores; wet liners insert with a light force fit and come into direct contact with the coolant. Seals confine the coolant to areas adjacent to ring travel (Figure 8-30). Wet liners simplify foundry work and give better control of water-jacket dimensions. The interface between dry liners and cylinder bores erects a minor, but real, thermal barrier between combustion and coolant. However, the liner is a structural member and eliminates the possibility of coolant leaks into the crankcase or back into the combustion chamber. Liners of either type stand proud of the deck, a practice that gives additional compression to the head gasket in a critical area and, at the same time prestresses the liner (Figure 8-31). Any liner that does not meet specification must be replaced.

This measurement taken at several points around the circumference of the liner, should be made during initial teardown and after replacement liners are installed. Figure 8-32 illustrates the liner hold-down hardware used in machine shops. Clamps are normally fabricated on site from cold-rolled steel, fitted with hardened washers, and pulled down securely. This technique works for dry liners and wet liners with radial (O-ring type) seals. Liner counterbores—the ledges that establish liner height—can be remachined and shimmed when necessary. The more usual procedure is to shim the liner shoulder, as shown in the illustration.

Some wet liners fit so loosely that seal resiliency can affect liner height when the cylinder head is removed. John Deere and other manufacturers that use this type of liner provide detailed loading instructions. Neither do engines fitted with these liners

Cylinder
liner

Crevice
seal

D-ring
seals

Integral
coolant
manifold

Rear main
bearing cap

Oil
galleries

8-30 Cylinder liner—wet type. Detroit Diesel

tolerate moving the crankshaft with the head detached. If this is done, piston-ring friction will raise the liners and possibly damage the O-ring seals.

Boring

Ascertain the amount of bore wear. Most wear occurs near the top of the cylinder at the extreme end of ring travel. This wear is caused by local oil starvation and

Outer flange

Protrusion

$\dfrac{0.002 \text{ in.}}{0.005 \text{ in.}}$

Sleeve

Shim (use only if
necessary)

Crankcase

8-31 Liner protrusion measurement is taken at several places around the circumference of the part and averaged. International Harvester

Holding adapter

Sleeve

Capscrew

8-32 Sleeve hold-down hardware. International Harvester

combustion-related acids. Pronounced wear, sometimes taking the form of scuffing, can occur at right angles to the crankshaft centerline on the bore surface that absorbs piston angular thrust. An engine that turns clockwise when viewed from the front will show more wear on the right side of its cylinders than on the left. In addition, axial thrust forces can generate wear in bore areas adjacent to the crankshaft centerline (Figure 8-33). These forces account for most of the taper and eccentricity exhibited by worn cylinders. Block distortion accounts for the rest.

Some idea of bore condition can be had by inserting a ring into the cylinder with the flat end of a piston. The difference in ring end gap between the upper and lower portions of the cylinder, as determined with a feeler gauge, roughly corresponds to cylinder wear. However, such techniques do not substitute for repeated and averaged measurements with a cylinder bore gauge (Figure 8-34).

Study the surface of the bore under a strong light. Deep vertical scratches usually indicate that the air filter has at one time failed. The causes of more serious

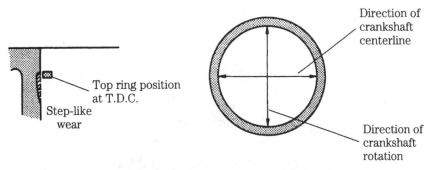

Top ring position
at T.D.C.

Step-like
wear

Direction of
crankshaft
centerline

Direction of
crankshaft
rotation

8-33 Areas of accelerated cylinder wear. Yanmar Diesel Engine Co. Ltd.

8-34 Using a cylinder gauge.
Chrysler Corp.

damage—erosion from contact with fuel spray, galling from lack of lubrication, rips and tears from ring, ring land, or wrist pin lock failure—will be painfully obvious.

Integral or dry-sleeve bores can be overbored and fitted with correspondingly oversized pistons. When the bore limit is reached, replacement sleeves can be fitted to either type, although the work is considerably easier on an engine that was originally sleeved.

The common practice is to use a boring bar for the initial cuts and finish to size with a hone, preferably an automatic, self-lubricating hone such as the Sunnen CV. The finish will approximate that achieved by the original equipment manufacturer (OEM).

The accuracy of the job can only be as good as the datum—the reference point from which all dimensions, including the bore-to-crankshaft relationship, are taken. Most boring bars index to the deck, on the assumption that the deck is parallel to the main bearing centerline. This is a large assumption. The better tools index to the main bearing saddles. The cylinder with most wear is bored first; how much metal must be removed to clean up this cylinder determines the bore oversize for the engine.

Oversized piston and ring assemblies are normally supplied in increments of 0.010 in. (or 0.25 mm) to the overbore limit. A few manufacturers offer 0.05 in. over pistons for slightly worn cylinders. Occasionally one runs across a 0.015-in. piston and ring set.

The overbore limit varies with sleeve thickness, cylinder spacing (too much overbore compromises the head gasket in the critical "bridge" area between cylinders), and the thickness of the water jacket for engines with integral bores. Jacket thickness depends, in great part, on the quality control exercised by the foundry. Some blocks and whole families of engines have fairly uniform jacket thickness; others are subject to core shift and the unwary machinist can strike water. A final consideration, not of real concern unless the engine is really "hogged out" beyond continence, is obtaining a matching head gasket. The gasket must not be allowed to overhang the bore.

Of course, it is always possible to replace dry liners and to install liners in worn integral bores. The latter operation can be expensive, and most operators would be advised to invest in another block.

Dry sleeves are, by definition, difficult to move; wet sleeves can also stick and some have more propensity for this than others. In short, liner removal and installation tools must be used. Figure 8-35 illustrates a typical combination tool. This or a similar tool must be used for extraction, but a press and a stepped pilot—one diameter matching

1 Bolt
2 Bracket
3 Patch plate
4 Cylinder liner
5 Packing
6 Bolt
7 Cylinder block
8 Bolt

8-35 Cylinder liner removal and installation tool. Marine Engine Div., Chrysler Corp.

liner inner diameter (ID), the other, liner outer diameter (OD)—is the better choice for installation.

Liner bores and counterbores (the ledge upon which the liner seats) require careful measurement and inspection. Out-of-spec counterbores can sometimes be machined and restored to height with shims, Cummins' fashion.

Wet-liner seals must be lubricated just prior to installation to control swelling. Seal contact areas on the water-jacket ID should be cleaned and inspected under a light. Some machinists oil dry sleeves; others argue against the practice.

Some sleeves, wet or dry, tend to crack at the counterbore area after a few hours of operation. The most common causes have to do with chamfers, either the chamfer on the counterbore or the chamfer on the liner installation pilot.

Detroit Diesel reboring

These two-cycle engines require some special instruction. Many Series 71 engines were cast in aluminum. Early production inserts were a slip fit in the counterbore; current standards call for the liner to be pressed in. In any event, the block should be heated to between 160° and 180°F in a water tank. Immerse the block for at least 20 minutes.

Counterbore misalignment can affect any engine, although mechanics generally believe that the aluminum block is particularly susceptible to it. You will be able to detect misalignment by the presence of bright areas on the outside circumference of the old liner. The marks will be in pairs—one on the upper half of the liner and the other diagonally across from it, on the lower half. The counterbore should be miked to check for taper and out-of-roundness.

Small imperfections—but not misalignment between the upper and lower deck—can be cleaned up with a hone. Otherwise, the counterbore will have to be machined. Torque the main-bearing caps. Oversize liners are available from the OEM and from outside suppliers such as Sealed Power Corporation.

Liners on early engines projected 0.002–0.006 in. above the block (Figure 8-36). These low-block engines used a conventional head gasket and shims under the liner

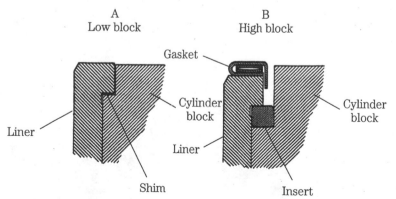

8-36 Detroit Diesel liner arrangements. Sealed Power Corp.

to obtain flange projection. Late-model high-block engines use an insert below the liner. Narrower-than-stock inserts are available to compensate for metal removed from the fire deck, and in various oversized diameters to accommodate larger liners. A 0.002-in. shim is also available for installation under the insert.

Note: Inserts can become damaged in service and can contribute to upper liner breakage.

Series-53 engines employed a wet liner. The upper portion of the liner is surrounded by coolant and sealed with red silicone seals in grooves on the block. Early engines had seals at the top and above the ports. Late-model engines dispensed with the lower seal. A second groove was machined at the top of the cylinder to be used in the event of damage to the original.

The seals must be lubricated to allow the liners to pass over them. Do not pre-soak the seals, because silicone expands when saturated with most lubricants. The swelling tendency is pronounced if petroleum products are used. Lubricate just prior to assembly with silicone spray, animal fat, green soap, or hydraulic brake fluid. Carefully lower the liners into the counterbores, without twisting the seals or displacing them from their grooves.

The eccentricity (out-of-roundness and taper) must be measured before final assembly. On the 110 series you are allowed 0.0015 in. eccentricity. The 53 and 71 engines will tolerate 0.002 in. Eccentricity can often be corrected by removing the liners and rotating them 90° in the counterbores. Do not move the inserts in this operation.

Honing

Honing is used to bring rebored cylinders to size and to remove small imperfections and glaze in used cylinders. Glaze is the hard surface layer of compacted iron crystals formed by the rubbing action of the rings. Most engine manufacturers recommend that the glaze be broken to add in ring seating and to remove the ridge that forms at the upper limit of ring travel. The Perfect Circle people suggest that honing can be skipped if the cylinder is in good shape.

The pattern should be diamond-shaped, as shown in Figure 8-37, with a 22-32 intersection degree at the horizontal centerline. The cut should be uniform in both directions, without torn or folded metal, leaving a surface free of burnish and imbedded stone particles. These requirements are relatively easy to meet if you have access to an automatic honing machine. However, satisfactory work can be done with a fixed-adjustment hone turned by a drill press or portable drill motor.

8-37 Preferred crosshatch pattern.

The hone must be parallel to the bore axis. Liners can he held in scrap cylinder blocks or in wood jigs. The spindle speed must be kept low—a requirement that makes it impossible to use a ¼ in. utility drill motor. Suggested speeds are shown below.

Bore diameter (inches)	Spindle (rpm)	Speed strokes (per minute)
2	380	140
3	260	83
4	190	70
5	155	56

Move the hone up and down the bore in smooth oscillations. Do not let the tool pause at the end of the stroke, but reverse it rapidly. Excessive pressure will load the stone with fragments, dulling it and scratching the bore. Flood the stones with an approved lubricant (such as mineral oil), which meets specification 45 SUV at 100°F.

Stone choice is in part determined by the ring material. Most engines respond best to 220–280 grit silicon carbide with code J or K hardness.

Cleaning the bore is a chore that is seldom done correctly. Never use a solvent on a honed bore. The solvent will float the silicon carbide particles into the iron, where they will remain. Instead, use hot water and detergent. Scrub the bore until the suds remain white. Then rinse and wipe dry with paper towels. The bore can be considered "sanitary" when there is no discoloration of the towel. Oil immediately.

Piston rings

Piston rings are primarily seals to prevent compression, combustion, and exhaust gases from entering the crankcase. The principle employed is a kind of mechanical jujitsu—pressure above the ring is conducted behind it to spread the ring open against the cylinder wall. The greater the pressure above the ring, the more tightly the ring wedges against the wall (Figure 8-38).

The rings also lubricate the cylinder walls. The oil control ring distributes a film of oil over the walls, providing piston and ring lubrication. One or more scraper

Gas pressure

8-38 A compression ring is a dynamic seal, expanding under the effects of gas pressure.

rings control the thickness of the film, reducing chamber deposits and oil consumption. In addition to sealing and lubrication, the ring belt is the main heat path from the piston to the relatively cool cylinder.

Rings are almost always cast iron, although steel rings are used in some extreme-pressure situations. Cast iron is one of the very few metals that tolerate rubbing contact with the same material. Until a few years ago rings were finished as cast.

Today almost all compression rings are flashed with a light (0.004 in.) coating of chrome. Besides being extremely hard and thus giving good wear resistance, chrome develops a pattern of microscopic cracks in service. These cracks, typically accounting for 2% of the ring's surface, serve as oil reservoirs and help to prevent scuffing. A newer development is to *fill,* or *channel,* the upper compression ring with molybdenum. The outer diameter of the ring is grooved and the moly sprayed on with a hot-plasma or other bonding process. Besides having a very low coefficient of friction and a very high melting temperature, moly gives a piston ring surface that is 15–30% void. It retains more oil than chrome-faced rings and is, at least in theory, more resistant to scuffing.

Rings traditionally have been divided into three types, according to function. Counting from the top of the piston, the first and second rings are *compression* rings, whose task is to control blowby. The middle ring is the *scraper,* which keeps excess oil from the combustion space. The last ring is the *oil* ring, which is serrated to deliver oil to the bore.

This rather neat classification has become increasingly ambiguous with the development of multipurpose ring profiles and the consequent reduction in the number of rings fitted to a piston. Five- and six-ring pistons have given way to three- and four-ring pistons on many of the smaller engines. The function of the middle rings is split between gas sealing and oil control. The lower rings, while primarily operating as cylinder oilers, have some gas-seating responsibilities. Design has become quite subtle, and it is difficult for the uninitiated to distinguish between *compression* and *scraper* rings.

The drawing in Figure 8-39 illustrates the ring profiles used on GM Bedford engines. Note the differences in profile among the three. These profiles are typical, but by no means, universal. The Sealed Power Corporation offers several hundred in stock and will produce others on special order.

What this means to the mechanic is that he or she must be very careful when installing rings. Most have a definite *up* and *down,* which might or might not be indicated on the ring. Usually the top side is stamped with some special letter code. Great care must be exercised not to install the rings in the wrong sequence. New rings are packaged in individual containers or in groups that are clearly marked 1 (for first compression), 2, and so forth. Reusable rings should be taken off the piston and placed on a board in the assembly sequence.

Ring wear

The first sign of ring wear is excessive oil consumption, signaled by blue smoke. But before you blame the rings, you should check the bearing clearances at the main

C$_L$ connecting
rod

C$_L$ combustion
chamber

Oil bleed hole

8-39 Ring configuration and nomenclature. GM Bedford Diesel

and crankpin journals. Bearings worn to twice normal clearance will throw off five times the normal quantity of oil on the cylinder walls. You can make a direct evaluation of oil spill by pressurizing the lubrication system. If appreciable amounts of oil are getting by the rings, the carbon pattern on the piston will be chipped and washed at the edges of the crown.

Check the rings for sticking in their grooves (this can be done on two-cycles from the air box), breaks, and scuffing. The latter is by far the most common malady, and results from tiny fusion welds between the ring material and cylinder walls. Basically it can be traced to lack of lubrication, but the exact cause might require the deductive talents of a Sherlock Holmes. Engineers at Sealed Power suggest these possibilities

Symptom	Possible cause
Overheating	Clogged, restricted, sealed cooling system
	Defective thermostat or shutters
	Loss of coolant
	Detonation
Lubrication failure	Worn main bearings
	Oil pump failure
	Engine lugging under load
	Extensive idle
	Fuel wash on upper-cylinder bores
	Water in oil
	Low oil Level
	Failure pressurize oil system after rebuild
Wrong cylinder finish	Low crosshatch finish
	Failure to hone after reboring
Insufficient clearance	Inadequate bearing clearance at either end of the rod
	Improper ring size
	Cylinder sleeve distortion

Usually insufficient bearing clearance, complicated by a poor fit in the block counterbore, results in overheating. The fundamental cause is often poor torque procedures, or improperly installed sealing rings on wet-sleeved engines. A rolled or twisted sealing ring can distort the sleeve.

Ring breakage is due to abnormal loading or localized stresses. It can be traced to

- Ring sticking that overstresses the free end of the ring.
- Detonation caused by overly liberal use of starting fluid, dribbling injectors and out-of-time fuel delivery.
- Overstressing the ring on installation. Usually the ring breaks directly across from the gap.
- Excessively worn grooves, which allow the ring to flex and flutter.
- Ring striking the ridge at the top of the bore. The mechanic is at fault because this ridge should have been removed.

The last point—involving blame—can be sticky in a shop situation. Mechanics make mistakes the same as everyone else, and the number of mistakes is in part, a function of the complexity of the repair. Few people can overhaul a machine as complicated as a multicylinder engine without making some small error. Assessing blame, if only to correct the situation, is sometimes complicated by having the mechanic who built the engine tear it down. But careful examination of the parts usually points to the fault. For example, rings that have been fitted upside down show reversed wear patterns. Rings that have broken in service are worn on either side of the break, from contact with the cylinder walls. Ring that have been broken during removal show sharp crystalline breaks without local wear spots.

Pistons

Large diesel engines often employ cast-iron or steel pistons; smaller, high-speed engines generally use aluminum castings or forgings. That such pistons survive combustion temperatures that can reach 4500°F and cylinder pressures that, highly supercharged engines, can exceed 2000 psi, is a triumph of engineering over materials. Aluminum has a melting point of 1220°F and rapidly loses strength as this temperature is approached.

Construction

The heavy construction of these pistons—diesel pistons typically weigh half as much again as equivalent SI engine pistons-provides mechanical strength and the heat conductivity necessary to keep the piston crown at about 500°F. The standard practice is to direct a stream of oil to the underside of the crown, usually by means of spray jets. Some designers go a step further, and insulate the crown, which is relatively easy to cool, from the ring belt and skirt. Turn ahead to Figure 8-48 for an illustration of a heat dam. The cavity above the piston pin slows heat transfer by reducing piston wall thickness. Another approach is to lengthen the thermal path by grooving the area above the ring belt (Figure 8-40). Yet another approach is to apply a thermal coating to the upper side of the piston crown, thus confining heat to the combustion chamber (Figure 8-41).

The skirts to aluminum pistons run hotter than cast iron and have a coefficient of expansion that is about twice that of iron. Consequently, light-metal pistons are assembled with fairly generous bore clearances to compensate for thermal expansion, and might be noisy upon starting. Semi-exotic alloys, such as Lo-Ex, or cast-in steel struts, help control expansion and knocking. It is interesting that one of the first experimenters with aluminum pistons, Harvey Marmon, found it necessary to sheathe the piston skirt in an iron "sock."

8-40 Fire grooves form thermal barriers between the ring band and combustion chamber. Yanmar Diesel Engine Co. Ltd.

Alumite treatment

8-41 Ford 2.3L pistons are insulated with a plasma coating and relieved at areas adjacent to the pins. Note the letter "F" embossed on the relief, which should face the front of the engine when the piston is installed.

Front

Most alloy pistons are cam-ground; when cold the skirts are ovoid, with the long dimension across the thrust faces. As the piston heats and expands, it becomes circular, filling the bore. Other ways of coping with thermal expansion are progressively to reduce piston diameter above the pin and to relieve, or cut back, the skirts in the pin area (also shown in Figure 8-41). The heavy struts that support the pin bosses transfer heat from the underside of the crown to the skirts.

Critical wearing points include the thrust faces and the walls of the ring grooves. With the exception of certain two-cycle applications, rings are designed to rotate in their grooves and, according to one researcher, reach speeds of about 100 rpm. Rotation is the primary defense against varnish buildup and consequent ring sticking. But it also wears "steps" into the grooves. Piston thrust faces are also subject to rapid wear.

Some manufacturers run the compression ring against a steel insert, cast integrally with the piston. Another approach is to substitute a long-wearing eutectic alloy for the ASE 334 or 335 usually specified. Eutectic alloys consist of clusters of hard silicon crystals distributed throughout an aluminum matrix. As the aluminum wears, silicon—one of the hardest materials known—emerges as the bearing surface. The same mechanism rapidly dulls cutting tools, which is why the cost of eutectic pistons approaches the cost of forgings.

Forged pistons are primarily an aftermarket item, used as a last resort in highly supercharged engines when castings have failed. Forging eliminates voids in the metal and compacts the grain structure at the crown, pin bosses, and ring lands (Figure 8-42). These pistons have superior hot strength characteristics but require generous running clearances. An engine with forged pistons will be heard from during cold starts.

8-42 Forged pistons gain strength from the uniform grain structure.
Sealed Power Corp.

A two-piece piston consists of a piston dome, or ring carrier, element and a skirt element (Figure 8-43). These parts pivot on the piston pin. Although other manufacturers use this form of the piston, the GM version first appeared on Electromotive railroad engines and, in 1971, replaced conventional trunktype pistons on turbocharged DDA Series 71 engines. It eventually found its way into several other DDA engines, including the Series 60.

Detroit Diesel describes this design as a "crosshead" piston. As the term is usually applied, it refers to a kind of articulated piston used on very large engines. A pivoted extension bar separates the upper piston element and the skirt, which rides against the engine frame. The crosshead isolates the cylinder bores from crankshaft-induced side forces. The DDA design does not relieve the bore of side forces, but it increases the bearing area of the pin and centralizes the load more directly on the conn rod for a reduction in bending forces. The connecting rod, illustrated in the following section, bolts to the underside of the pin, making the upper half of the pin available to support piston-dome thrust.

Failure modes

Piston failure is usually quite obvious. Wear should not be a serious consideration in low-hour engines, because the skirt areas are subject to relatively small forces and have the benefit of surplus lubrication. Excessive wear can be traced to dirty or improperly blended lube oil or inadequate air filtration. Poor cylinder finishing might also contribute to it. Piston collapse or shrinkage is usually due to overheating. If the problem shows itself in one or two cylinders, expect water-jacket stoppage or loose liners.

Combustion roughness or *detonation* damage begins by eroding the crown, usually near the edge. The erosion spreads and grows deeper until the piston "holes." Typically the piston will look as if it were struck by a high-velocity projectile. *Scuffing*

Fire ring

Compression ring

Oil control ring

Oil ring expander

Piston dome and bushing assembly

Piston skirt

Piston pin

8-43 Detroit Diesel articulated piston reduces bending forces on the piston pin.

and *scoring* (a scuff is a light score) might be confined to the thrust side of the piston. If this is the case, look for the following:

- Oil pump problems such as clogged screen, excessive internal clearances.
- Insufficient rod bearing clearances, which reduce throw-off, robbing the cylinders of oil.
- Lugging.

The probable causes of damage to both sides of the skirt include the ones just mentioned, plus the following:

- Low or dirty oil.
- Detonation.

- Overheating caused by cooling system failure.
- Coolant leakage into the cylinder.
- Inadequate piston clearance.

Scuffs or scores fanning out 45° on either side of the pinhole mean one of the following conditions:

- Pin fit problems such as insufficient bearing clearances at the small end of the rod or at the piston bosses.
- Pin boss damage (see below for installation procedures).

Ring land breakage can be caused by the following:

- Excessive use of starting fluid.
- Detonation.
- Improper ring installation during overhaul.
- Excessive side clearance between the ring and groove.
- Water in cylinder.

Free-floating pins sometimes float right past their lock rings and contact the cylinder walls. Several causes (listed below) have been isolated.

- Improper installation: Some mechanics force the lock rings beyond the elastic limit of the material. In a number of cases, it is possible to install lock rings by finger pressure alone.
- Improper piston alignment: This might be caused by a bent rod or inaccuracies at the crankshaft journal. Crank pins that are tapered or out of parallel with the main journals will give the piston a rocking motion that can dislodge the lock ring. Pounding becomes more serious if the small-end bushing is tight.
- Excessive crankshaft end play: Fore-and-aft play is transmitted to the lock rings and can pound the grooves open. Again, a too-tight fit at the connecting rod's small end will hasten piston failure.

Servicing

Used pistons can give reliable service in an otherwise rebuilt engine, but only after the most exhaustive scrutiny. Scrape and wire-brush carbon accumulations from the crown, but do not brush the piston flanks. Carbon above the compression ring and on the underside of the crown should be removed chemically. A notch, letter, arrow, or other symbol identifies the forward edge of the piston. These marks are usually stamped on the crown, on the pin boss relief (illustrated in Figure 8-41), or hidden under the skirt. Make note of the relationship between the leading edge of the piston and the numbered side of the connecting rod.

Lay out the rings in sequence, topsides up, on the bench. Spend some time "reading" the rings—the history of the upper engine is written on them, just as the crank bearings testify to events below. The orientation code—the word *Top* (T) or *Ober.* (O)—will be found on the upper sides of the compression and scraper rings, adjacent to the ring ends.

You might wonder why attention is given to these codes for piston assemblies that, at this stage, are of unknown quality, and for rings that will, in any event, be discarded. The purpose is to become familiar with the concept of orientation as it applies to the particular engine being serviced. New parts might not carry identical codes, but the relationship between coded parts will not change.

While still attached to the rod and before investing any more time in it, inspect the piston for obvious defects. Reject if the piston is fractured, deeply pitted, scored, or if it exhibits ring land damage (Figure 8-44). Cracks tend to develop where abrupt changes in cross section act as stress risers. Deep pits usually develop at the edges of the piston crown; scoring is most likely to develop on the thrust faces. Contact with the cylinder bore occurs at two areas, 90° from the piston-pin centerline. The major thrust face lies in the direction of the crankshaft rotation and is normally the first to score. (Viewed from the front of a clockwise-rotation engine, the major thrust face is on the right.) Machine marks, the light cross-hatching left by the cutting tool, should remain visible over most of the contact area. The next section describes the relationship of wear patterns to crankshaft and rod alignment.

If the piston passes this initial examination, make a micrometric measurement of skirt diameter across the thrust faces. Depending on the supplier, the measurement is made at the lower edge of the skirt, at the pin centerline, or at an arbitrary distance

8-44 Typical piston failure modes. ©1990 Deers & Co.

8-45 Piston measurement points. OD measurements may vary with manufacturers. Yanmar Diesel Engine Co. Ltd.

between these points (Figure 8-45). Piston OD subtracted from cylinder bore ID equals running clearance.

All traces of carbon in the ring grooves and oil spill ports must be removed. Whenever possible, this tedious task should be consigned to the machinist and thought about no more. Figure 8-46 illustrates a factory-supplied plug gauge used to

8-46 The best way to determine ring groove wear is with a factory-supplied go-nogo gauge. International Harvester.

Feeler gauge

Ring-to-groove clearance

8-47 An alternative method of determining groove wear is to measure the width relative to a new ring. Yanmar Diesel Engine Co. Ltd.

measure groove width; the next drawing, Figure 8-47, shows an alternate method using a *new* ring and a feeler gauge. The latter method is accurate, so long as the gauge bottoms against the back of the groove.

Most engines of this class employ "full-floating" piston pins that, at running temperatures, are free to oscillate on both the rod and piston (Figure 8-48). Pins secure laterally with one or another variety of snap rings, some of which are flattened on the inboard (or wearing) sides.

Protecting your eyes with safety glasses, disengage and withdraw the snap rings. Although mechanics generally press out (and sometimes hammer out) piston pins, these practices should be discouraged. Instead, take the time to heat the pistons, either with a heat gun or by immersion in warm (160°F) oil. Pins will almost fall out.

While the piston is still warm, check for bore integrity. Insert the pin from each side. If the pin binds at the center, the bore might be tapered; if the bore is misaligned, the pin will click or bind as it enters the far boss (Figure 8-49).

Other critical areas are illustrated in Figure 8-45. Measurements are to be made at room temperature. Damage to retainer rings or ring grooves suggests excessive crankshaft end play, connecting-rod misalignment, or crankpin taper. Slide forces generated by a badly worn crankpin or severely bent rod react against the snap rings. Rod or crankpin misalignment might also appear as localized pin and pin-bearing wear, a subject discussed in the following section.

First compression ring

Second compression ring

Oil ring

Piston

Circlip

Piston pin

Circlip

8-48 Like most manufacturers, Yanmar favors full-floating piston pins, pivoting on both the rod and piston, and secured laterally by snap rings ("circlips" in the drawing).

To install a piston on its connecting rod, warm the piston, oil the pin and pin bores, and, with the rod in its original orientation, slip the pin home with light thumb or palm pressure. Use *new* snap rings, compressing them no more than necessary. Verify that snap rings seat around their full diameters in the grooves.

A   B

8-49 This drawing, intended by John Deere to show how piston pin bearing checks (free-floating pins should insert with light thumb pressure; if the pin goes in easily from both sides but binds in the center, the problem might be tapered bores (A); if the pin clicks or binds when contacting the far boss, the piston is warped (B), also illustrated the cavity on the underside of the head that serves as a heat dam.

A few engines use pressed-in pins that lock on the piston with an interference fit. Support the piston on a padded V-block and press the pin in two stages: stop at the point of entry to the lower boss, relieve press force to allow the piston to regain shape, and press the pin home. If the pin is installed in one pressing, the lower boss might be shaved.

What remains is to establish the running clearance. Piston-to-bore clearance is fundamentally a matter of specification, but specifications are never so rigid that they cannot be bent. For example, one John Deere piston/liner combination requires a running clearance of 0.0034–0.0053 in., as measured at the bottom of the piston skirt. A high-volume rebuilder typically goes toward the outside limit, building clearance here and in the crankshaft bearings. A "loose" engine will be more likely to tolerate severe loads as delivered and without the benefit of a break-in period. A custom machinist, who works on one or two engines at a time, often goes in the other direction, aiming at the tightest clearance the factory allows. Besides being aesthetically more satisfying, "tight" engines tend to live long, quiet lives. Of course, such an engine must be carefully run in during the first hours of operation.

In an attempt to standardize the maintenance process, factory manuals include, wear limits for critical components. The concept of permissible wear is a value judgment, made in an engineering office remote from the world of mechanics, back-ordered parts, and budgetary restraints. Permissible piston-to-liner clearance for one Deere engine is 0.0060 in. Suppose that the running clearance is found to be 0.0055 in. Assuming that the wear is equally distributed over both parts, should the piston or liner or both be replaced? This is only one cylinder of six, none of them worn by identical amounts. Questions like this go beyond mechanics and depend for their answers on the politics of the situation, interpreted in light of the philosophy of maintenance—formal or informal—that characterizes the operation.

Connecting rods

No part is more critical than the connecting rod and none more conservatively designed (Figure 8-50).

Construction

Most diesel engines employ two-piece, H-section rods, heavily faired at the transitions between bearing sections, and forged from medium-carbon steel or from that ubiquitous alloy, SAE 4140. The small end is generally closed and fitted with a replaceable bushing (some automotive OEMs do not catalog replacement bushings, but a component machinist can work around that). The big end carries a two-piece precision insert bearing. A few rods terminate in a slipper, which bolts to the piston pin, as shown in Figure 8-51. Open construction makes the whole length of the pin available as a bearing surface.

Piston pin bushing

Connecting rod

Crank pin bushing

Connection rod bolt

8-50 Typical connecting rod and bearing assembly. A rifle-drilled oil passage indexes with ports in both the small-end bushing and upper big-end insert. Yanmar Diesel Engine Co. Ltd.

Connecting rod

Bearing cap

Bearing shells

8-51 Detroit Diesel slipper-type connecting rod, used with the articulated piston shown in Figure 8-44.

225

Whenever possible, the parting line between the rod shank and cap should be perpendicular to the crankshaft centerline. This configuration reduces side forces on the cap. However, the rod must be split at an angle to allow disassembly through the bore when crankpin diameter is large (as in Onan engines) or to prevent contact with other parts (Deere 6076).

Connecting rods have a definite orientation relative to the piston and to the cap. The former is a function of transverse oil ports, drilled in the rod shank; the latter reflects the way connecting rods are manufactured. The relatively small connecting rods that we are dealing with are forged in one piece. Then the cap is separated (sometimes merely by snapping it off), assembled, and honed to size. Reversing the cap or installing a cap from another rod destroys bearing circularity. Typically, a mismatched assembly overheats and seizes within a few crankshaft revolutions. Rod shanks and caps carry match marks, and both parts are identified by cylinder number (Figures 8-52 and 8-53). Errors can be eliminated by double- and triple-checking match mark alignment and cylinder numbers. Mechanics do well to observe the rule that no more than one rod cap should be disassembled at a time.

Service

The number stamped on the rod shank and cap should correspond to the cylinder number. Sometimes these numbers are scrambled or missing, and the mechanic must supply them. Stamp the correct numbers on the pads provided and, to prevent confusion, deface the originals.

Mike rod journals at several places across the diameter, repeating the measurements on both sides to detect taper (Figure 8-54). Inertial forces tend to stretch the rod cap and pinch the ends together. When this condition is present, the machinist

Alignment mark
(casting mark)

8-52 Most rods and caps carry cast match marks as shown on this Yanmar assembly. Cylinder numbers will be stamped on the other side of the rod.

8-53 Rod assembly for No. 1 cylinder on a Detroit Diesel engine.

should mill a few thousandths off the cap interface and, with the cap assembled and torqued, machine the journal to size. Typically, this is done with a reamer, although automatic honing machines, such as the Sunnen or the Danish-made AMC, produce a finer, more consistent surface.

The machinist should check the alignment of each connecting rod with a fixture similar to the one shown in Figure 8-55. Even so, these matters should not be entirely left to the discretion of the machinist. It is always prudent to check the work against the testimony of the engine. Rod and crankpin misalignment produces telltale wear patterns on piston thrust faces, on pin bores, and on the retainers and retainer grooves.

8-54 Journal measurement points. Marine Engine Div., Chrysler Corp.

8-55 Rod alignment fixture used to detect bending and torsional misalignment. Yanmar Diesel Engine Co. Ltd.

A bent rod tilts the piston, localizing bearing wear at the points shown in Figure 8-56, This condition might also be reflected by an hourglass-shaped wear pattern on the thrust faces, as illustrated in Figure 8-57. A twisted rod imparts a rocking motion to the piston, concentrating wear on the ring band and skirt edges (Figure 8-58). Crankshaft taper generates thrust, which might scuff the skirt and could drive the pin through one or the other retainer.

Bent or twisted rods generally can be straightened, although some manufacturers warn against the practice.

Most rod bolts can be reused, if visual and magnetic-particle testing fails to reveal any flaws. Some recently developed torque-to-yield bolts can also be reused;

8-56 A bent rod concentrates bearing wear at the points shown. Ford Motor Co.

8-57 Another effect of loss of rod parallelism is to tilt the piston and, in extreme cases, produce the wear pattern shown on the left. Sealed Power Corp.

earlier types were sacrificial. Rod-bolt nuts should be replaced, regardless of the fastener type.

Used rods should be Magnafluxed. Interpretation of the crack structure thus revealed requires some judgment. Any rod that has seen service will develop cracks. One must distinguish between inconsequential surface flaws and cracks that can lead to structural failure. The drawing in Figure 8-59 can serve as a guide.

8-58 A twisted connecting rod causes the piston to rock, a condition that can be signaled by an elliptical wear pattern. Sealed Power Corp.

Start of fatigue crack resulting from overloading (due to hydrostatic lock). Do not attempt to salvage (this type of indication is not visible with bushings in place)

Do not use or attempt to salvage rods with indications over 0.005 in. deep extending over edges of H-section on both sides of flange (shaded areas are most highly stressed)

Section A-A

Section B-B

Example of an indication following longitudinal forging flow lines

Noncritical area

Critical area acceptable limits

Noncritical area

Indications in noncritical areas are acceptable unless they can be observed as obvious cracks without magnetic inspection

Example of a transverse indication that does not follow longitudinal forging flow lines can be either a forging lap, heat treat crack, or start of a fatigue crack

Longitudinal indications
Following forged flow lines are usually seams and are not considered harmful if less than 1/32 in. deep

Poor practice   Good practice

Grinding notes
Care should be taken in grinding out indications to assure proper blending of ground area into unground surface so as to form a smooth contour

Transverse indications (across flow lines) Having a maximum length of 1/2 in. which can be removed by grinding no deeper than 1/64 in. are acceptable after their **complete removal.** An exception to this is a rod having an indication which extends over the edge of H-section and is present on both sides of the flange in this case. Maximum allowable depth is 0.005 in. (see section A-A).

8-59 Interpreting Magnaflux indications. Detroit Diesel

In general, longitudinal cracks are not serious enough unless they are 1/32 in. deep, which can be determined by grinding at the center of the crack. Transverse cracks are causes of concern because they can be the first indication of fatigue. If the cracks do not extend over the edges of the H-section, are no more than 1/2 in. long, and less than 1/64 in. deep, they can be ground and feathered. Cracks over the H-section can be removed if 0.005 in. deep or less. Cracks in the small end are cause for rejection.

Small-end bushings are pressed into place with reference to the oil port and are finish-reamed. Loose bushings are a sign that the rod has overheated, and they can cause a major failure by turning and blocking the oil port.

What has been described are standard, industry-wide practices. One can go much further in the quest for a more perfect, less problematic engine. For example, it is good practice to match the weight of piston and rod assemblies to within 10 g or so—even when parts are not to be sent out for balancing. Surplus weight can be ground from the inner edges of the piston skirts. Some mechanics assemble one piston and rod to several crankpins. Differences in deck height between No. 1 and the last cylinder of the bank indicate an alignment problem, either between crankshaft throws or between the crankshaft centerline and deck.

Crankshafts

At this point we have progressed to the heart of the engine, the place where its durability will be finally established.

Construction

Crankshafts for the class of engine under discussion are, for the most part, steel forgings. Materials range from ordinary carbon steels to expensive alloys such as chrome-moly SAE 4140. A number of automotive engines and lightly stressed stationary power plants get by with cast-iron shafts, usually recognized by sharp, well-defined parting lines on the webs and by cored crankpins. Forging blurs the parting lines and mandates solid crankpins, which, however, are drilled for lubrication.

Most crank journals are induction-hardened, a process that leaves a soft, fatigue-resistant core under a hardened "skin." Hardness averages about 55 on the Rockwell scale and extends to a depth of between 0.020–0.060 in. in order to accommodate regrinding. But a cautious machinist will test crankshaft hardness after removing any amount of metal.

A few extreme-duty crankshafts are hardened by a proprietary process known as *Tufftriding*. Wearing quality is comparable to that provided by chrome plating, but unlike chromium, the process does not adversely affect the fatigue life of the shaft.

Service

Remove the crankshaft from the block—a hoist will be required for the heavier shafts—and make this series of preliminary inspections:

- If a main bearing cap or rod has turned blue, discard the shaft, together with the associated cap or rod. The metallurgical changes that have occurred are irreversible. By the same token, be very leery of a crankshaft that is known to have suffered a harmonic balancer failure.
- Check the fit of a new key in the accessory-drive keyway. There should be no perceptible wobble. A competent machinist can rework worn keyways and, if necessary, save the shaft by fabricating an oversized key.

8-60 Crankshaft measurement points. Marine Engine Div., Chrysler Corp.

- Check the timing-gear teeth for wear and chipping. Magnetic-particle testing can be of some value when applied to the hub area, but cannot detect the subsurface cracks that signal incipient gear-tooth failure. Timing gear sprockets and related hardware should be replaced as a routine precaution during major repairs.
- Mike the journals and pins as shown in Figure 8-60. Compare taper, out-of-roundness, and diameter against factory wear limits.
- If corrective machining does not appear necessary, remove with crocus cloth all light scratches and the superficial ridging left by bearing oil grooves. Tear off a strip of crocus cloth long enough to encircle the journal. Wrap the cloth with a leather thong, crossing the ends, to apply force evenly over the whole diameter of the journal. Work the crocus cloth vigorously, stopping at intervals to check progress. An Armstrong grinder works surprisingly fast.
- Using an EZ-Out or hex wrench, remove the plugs capping the oil passages. Back-drilled sections of these passages serve as chip catchers, and must be thoroughly cleaned. Compressed air, shown in Figure 8-61, helps but is no substitute for rifle-bore brushes, solvent, and elbow grease. Clean the plugs, seal with Loctite, and assemble.
- Check trueness with one or, preferably, tandem dial indicators while rotating the crank in precision V-blocks (Figure 8-62).

8-61 Blind runs in the crankshaft oiling circuit must be opened, mechanically cleaned, and resealed. Lombardini

Dial gauge

Journal

8-62 Checking crankshaft straightness using the center bearing as the referent. Yanmar Diesel Engine Co. Ltd.

Flaw testing

Flaw testing is generally done with the Magnaflux process, although some shops prefer to use the fluorescent-particle method. Both function on the principle that cracks in the surface of the crankshaft take on magnetic polarity when the crank is put in a magnetic field. Iron particles adhere to the edges of these cracks, making them visible. The fluorescent particle method is particularly sensitive because the metal particles fluoresce and glow under black light.

Most cracks are of little concern because the shaft is loaded only at the points indicated in Figure 8-63. The strength of the shaft is impaired by crack formations which follow these stresses, as shown in Figure 8-64. These cracks radiate out at 45° to the crank centerline and will eventually result in a complete break.

Abnormal bending forces are generated by main-bearing bore misalignment, improperly fitted bearings, loose main-bearing caps, unbalanced pulleys, or over-tightened belts. Cracks caused by bending start at the crankpin fillet and progress diagonally across.

8-63 Most crankshaft loads pass through the webs. Detroit Diesel

8-64 Typical fatigue crack patterns, all of them Critical. Detroit Diesel.

The distribution of cracks caused by torsional (or twisting) forces is the same as for bending forces. All crankshafts have a natural period of torsional vibration, which is influenced by the length/diameter ratio of the crank, the overlap between crankpins and main journals, and the kind of material used. Engineers are careful to design the crankshaft so that its natural periodicity occurs at a much higher speed than the engine is capable of turning. However, a loose flywheel or vibration damper can cause the crank to wind and unwind like a giant spring. Unusual loads, especially when present in conjunction with a maladjusted governor, can also cause torsional damage.

Crankshaft grinding

Bearings are available in small oversizes (0.001 and 0.002 in.) to compensate for wear. The first regrind is 0.010 in. Some crankshafts will tolerate as much as 0.040 in., although the heat treatment is endangered at this depth. The crankpin and main-journal fillets deserve special attention. Flat fillets invite trouble, because they act as stress risers. Gently radius the fillets as shown in the left drawing in Figure 8-65.

All journals and pins should be ground, even if only one has failed. Use plenty of lubricant to reduce the possibility of burning the journal. Radius the oil holes with a stone and check the crankshaft again for flaws with one of the magnetic particle methods.

Surface hardness should be checked after machining and before the crank receives final polishing. Perhaps the quickest and surest way to do this is to use Tarasov etch. Clean the shaft with scouring powder or a good commercial solvent. Wash thoroughly and rinse with alcohol. Apply etching solution No. 1 (a solution of 4 parts nitric acid in 96 parts water). It is important to pour the acid into the water, not vice versa.

3.5 R

OK   X   X   X

8-65 Fillet profiles. Sharp edges, flat surfaces, and overly wide profiles should be avoided. Rolled, as opposed to ground, fillets extend fatigue life by as much as 60%, but few shops have the necessary equipment. Detroit Diesel

Rinse with clean water and dry. If you use compressed air, see that the system filter traps are clean. Apply solution No. 2, which consists of 2 parts hydrochloric acid in 98 parts acetone. Acetone is highly flammable and has a sharp odor that can produce dizziness or other unpleasant reactions when used in unventilated areas, so allow yourself plenty of breathing room.

The shaft will go through a color change if it has been burned. Areas that have been hardened by excessive heat will appear white; annealed areas turn black or dark gray. Unaffected areas are neutral gray. If any color other than gray is present, the shaft should be scrapped and the machinist should try again, this time with a softer wheel, a slower feed rate, or a higher work spindle speed. Some experiment-ation might be necessary to find a combination that works.

Cranks that have been Tufftrided must be treated after grinding to restore full hardness. One test is chemical: a 10% solution of copper ammonium chloride and water applied to the crankshaft reacts almost immediately by turning brown if a traditional heat treatment has been applied. There will be no reaction in 10 seconds if the crankshaft is Tufftrided. Another test is mechanical: Tufftriding is applied to the whole crankshaft, not just to the bearing journals. If a file skates ineffectually over the webs without cutting, one can assume that the shaft is Tufftrided.

Any heavy-duty crankshaft, forged or cast, can benefit from Tufftriding, if the appropriate bearing material is specified. This treatment is especially beneficial when the manufacturer has neglected to treat journal fillets. The abrupt change in hardness acts as a stress riser. Treatment should be preceded by heating the crankshaft for several hours at a temperature above 1060°F. It might be necessary to re-straighten the crankshaft.

Camshafts and related parts

Inspect the accessory drive gear train for tooth damage and lash. Crank and camshaft gears are pressed on their shafts and further secured by pins, keys, or bolts.

Cam and balance shafts are supported on bushings, which require a special tool, shown in Figure 8-66, to extract and install. Normally, the machinist ser-vices in-block bushings, removing the old bushings and installing replacements that, depending upon engine type, might require finishing. The OEM practice of leaving bushing bores unfinished and reaming installed bushings to size complicates the rebuilding process. It is also possible for bushings to spin in their bosses, a condition that can be corrected by sleeving.

Camshaft lobes are the most heavily loaded parts of the engine, with unit pres-sures in excess of 50,000 psi at idle. Valve lifters, whether hydraulic or mechanical, are normally offset relatively to the lobes, so that contact occurs over about half of the cam surface. This offset, together with a barely visible convexity ground on the lifter foot tends to turn the lifter as it reciprocates. As the parts wear, the contact area increases and eventually spreads over the whole width of the lobe. Thus, if you find a camshaft with one or more "widetrack" lobes, scrap the cam, together with the complete lifter set. Although the condition is rare in diesel engines, overspeeding and consequent valve float might batter the flat toes of the lobes into points. Figure 8-67,

Guide (rear) Camshaft bushing Replace bar Cylinder block Bushing tool Guide (front)

Bushing tool

8-66 Camshaft bearing tool. Marine Engine Div., Chrysler Corp.

which illustrates lobe wear measurements, inadvertently illustrates this condition. The same two-point measurements should be made on the journals.

Normally, hydraulic valve lifters are replaced during a rebuild; attempting to clean used lifters is a monumental waste of time. This means that the camshaft must also be replaced, because new lifters will not live with a used camshaft (and vice versa). Assemble with the special cam lubricant provided, smearing the grease over each lobe and lifter foot. Oil can be used on the journals.

Replacement hydraulic lifters normally are charged with oil as received, and ample·time must be allowed for bleed when screwing down the rocker arms on OHV engines. Some valves will be open; if the rockers are tightened too quickly, pushrods will bend. As mentioned earlier, the low bleed-down rates characteristic of some lifters make rocker assembly an exercise in patience.

Break in the camshaft exactly per maker's instructions. Otherwise, it will score.

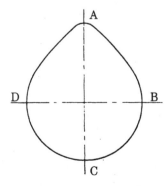

8-67 Cam lobe measurement points. Navista.

Harmonic balancers

The harmonic balancer, or vibration dampener, mounts on the front of the crankshaft where it muffles torsional vibration. Power comes to the crankshaft as a series of impulses that cause the shaft to twist, first in the direction of rotation and then against normal rotation. The shaft winds and unwinds from a node point near the flywheel. This movement can, unless dampened, quickly break the shaft.

Most harmonic balancers consist of an outer, or driven, ring bonded by means of rubber pads to the hub, which keys to the crankshaft. The rubber medium dampens crankshaft accelerations and deceleration, transferring motion to the outer ring at average crankshaft velocity. Some balancers drive through silicon-based fluid that exerts the same braking effect.

It is difficult to test a harmonic balancer in a meaningful way. Rubberized balancers can be stressed in a press and the condition of the rubber observed. Cracks or separation of the bonded joint means that the unit should be replaced. Fluid-filled balancers are checked for external damage and fluid leaks. Some machinists equate noise when the balancer is rolled on edge with failure.

Detroit Diesel's advice is best: replace the balancer whenever the engine is rebuilt.

Crankshaft bearings

All modern engines are fitted with two-piece precision insert bearings at the crankpin, as illustrated in Figures 8-50 and 8-51. Most employ similar two-piece inserts, or shells, at the main journals, although full-circle bearings are appropriate for barrel-type crankcases. No bearing can be allowed to spin in its carrier.[1] Full-circle bearings pressed into their carriers or locked by pins or cap screws. Two-piece shells secure with a tab and gain additional resistance to spinning from residual tension. Bearing shells are slightly oversized, so that dimension A in Figure 8-68 is greater than dimension B. The difference is known as the *crush height.* Torquing the cap equalizes the diameters, forcing the inserts hard against their bosses.

Thrust bearings, usually in the form of flanges of main inserts, limit crankshaft fore-and-aft movement (Figure 8-69). So long as the crankpin is not tapered, connecting-rod thrust loads are insignificant and transfer through rubbing contact between the rod big end and the crankpin flanges.

Insert-type bearings are made up in layer-cake fashion of a steel backing and as many as five tiers of lining material. Although their numbers are dwindling, perhaps half of the high-speed engines sold in this country continue to use copper-lead bearings, the best known of which is the Clevite (now Michigan)-77. The facing surface of this bearing consists of 75% copper, 24% lead, and a 1% tin overplate. Daimler-Benz, Perkins, and several other manufacturers have followed the example of Caterpillar and specify aluminum-based bearings, which tolerate acidic oils better

---

[1]Henry Ford made an exception to the "no-spin" rule; his early V-8s used floating crankpin bearings, babbitted on both sides. This technique cut bearing speed in half, but introduced a complication in the form of big-end connecting rod wear.

Difference between
diameters A and B is
bearing crush C

Diameter A at right angles
to parting lines greater
than diameter B

With bearing cap
drawn up tight
diameters A and B are equal

8-68 Bearing crush. Navistar

than copper-lead types. One material commonly used is Alcoa 750, an alloy of almost 90%-pure aluminum, 6.5% tin, alloyed with zinc and trace amounts of silicon, nickel, and copper.

Normal bearing wear should not exceed 0.0005 in. per 1000 hours of continuous operation. The operative word is "continuous"—frequent startups, cold loads, and chronic overloads can subtract hundreds of hours from the expected life. In practice, main bearings outlast crankpin bearings by about three to one, although No. 1 main, which receives side loads from accessories, can fail early.

8-69 Thrust bearing configuration for two Yanmar engines. Other makes use two-piece thrust washers.

8-70 Saddle alignment is determined with precision straight-edge and feeler gauge. Ford Motor co.

According to Michigan Bearing, 43% of failings can be attributed to dirt, 15% to oil starvation, 13% to assembly error, and 10% to misalignment. Overloading, corrosion, and miscellaneous causes account for the remainder. In other words, the mechanic contributes to most failures, by either assembling dirt into the engine, reversing nonsymmetrical bearing shells or bearing caps, or failing to provide adequate lubrication during initial start-up. These matters are discussed in the next section.

Bearing alignment problems come about from improperly seated rod caps or from warped main-bearing saddles. The condition can sometimes be spotted during disassembly as accelerated wear on outboard or center mainbearing shells. The mechanic should verify alignment by spanning the saddles with a precision straight-edge (Figure 8-70). Loss of contact translates as block warp.

Bearing alignment can be restored, but at the cost of raising the crankshaft centerline a few thousandths of an inch. The effect of chain-driven camshafts is to retard valve timing by an almost imperceptible amount; but gear trains are not so forgiving and the machinist might be forced to resort to some fairly exotic (and expensive) techniques to maintain proper tooth contact. The "cleanest" solution is to resize the bearing bosses by a combination of metalizing and honing.

Assembly—major components

Building up the major components—crankshafts, pistons, and rods—is a special kind of activity, that can be neither forced nor hurried.

Although remote from our subject, a certain insight into it can be gained from the experience of U.S. Naval gunners almost a century ago. As the century closed, five warships fired at a hulk from a mile distance for 25 minutes and registered two hits. This was an average performance, achieved by taking approximate aim and waiting for the roll of the ship to bring the sights into alignment with the target. Meanwhile an

Englishman by the name of Timms had developed another system of aiming, which involved constant correction, in an attempt to keep the sights on the target however the ship moved. His explanations that the gunner would naturally adjust to ship's motion went unheeded by Navy brass until he got the ear of Teddy Roosevelt, who put him in charge of gunnery training. In 1910, the test described above was repeated, this time using the new method and only one ship. The number of hits doubled.

Timms scored by concentrating directly on the problem and not on some ritualized technique that was supposed to yield an automatic solution. There is a great deal of room in the mechanic's trade for this kind of pragmatic thinking. For example, the whole business of measurement could be reformed, beginning with the establishment of a datum line (such as the crankshaft centerline) from which deck flatness, cylinder bore centerlines, and bearing clearances would be generated. And much thought could be given to substituting a direct measurement of clamping force for the indirect system currently used. When a capscrew is tightened, about 90% of the torque load appears as friction between the underside of the fastener head and the work piece. A 5% miscalculation in friction values represents a 50% clamping force error.

Glaze breaking

Cast-iron bores develop a hard glaze in service, which most mechanics roughen with a hone to facilitate ring sealing. Consequently, worn cylinder bores are often honed when piston rings are replaced. Remachined cylinders are also honed, but this is usually the province of the machine shop. However, recently developed cylinder finishing techniques make honing problematic.

A brush hone, sized to cylinder diameter, such as the 120-grit BRM Flex-Hone pictured in Figure 8-71 is normally used. Obtaining the desired crosshatch pattern

8-71 Most mechanics break glaze with a brush, or ball, hone.

(shown earlier in Figure 8-37) depends on a four-sided relationship among the ring material, spindle speed, bore diameter, and stroke frequency. For what it is worth, chrome rings running in a 4-in. cylinder need the finish produced by a 280-grit stone, rotated 190 rpm and reciprocated 70 times a minute. Such precision is unobtainable, but a little practice will produce an acceptable surface.

It is important to use plenty of lubricant, such as mineral oil or PE-12, available in aerosol cans from specialty tool houses. Keep the hone moving—it cuts as it rotates—and do not pause at the ends of the strokes. Hone for 3 or 4 seconds and stop to inspect the bore. Crosshatched grooves should intersect at angles of 30–45° relative to the bore centerline. If the pattern appears flat, reduce the spindle speed or increase the stroke rate. The job should require no more than 15 or 20 seconds to complete. Withdraw the hone while it is still turning.

Few mechanics take the time to clean a bore properly. Solvents merely float abrasive hone fragments deeper into the metal. Detergent, hot water, and a scrub brush are the prescription. Scrub until the suds are white, and wipe dry with paper shop towels. Continue to scrub and wipe until the towels no longer discolor. According to TRW, half of early ring failures can be traced to abrasive particles left in the bore.

Laying the crank

This discussion applies to engines with two-piece main-bearing shells; full-circle bearings require a slightly different technique, as outlined in the appropriate shop manual.

1. Although it is late in the day, make one final inspection of the block. Critical areas include saddle alignment for two-piece GM block castings, cracks that are likely to be associated with saddle webbing and liner flanges, and main bearing cap-to-saddle fits. Most caps are buttressed laterally by interference fits with their saddles. Fretting on contact surfaces can, if not corrected, result in crankshaft failure. Some mechanics make it a practice to chase main-bearing bolt threads; if you opt to do this, use the best, most precise tap that money will buy.
2. Verify that main-bearing saddles are clean and dry.
3. Unwrap one set of main-bearing inserts. Bearing size will be stamped on the back of each insert. All machinists can tell stories about mislabeled bearings, but certainly it would give one pause if the crankshaft were turned 0.010 in. and the inserts were marked "Std."
4. Identify the upper insert, which always is ported for crankshaft lubrication. A few lower shells carry a superfluous port; most are blanked off and grooved.
5. Roll the upper insert into place on the web, leading with the locating tab, as shown in Figure 8-72. Make absolutely certain that the tab indexes with the web recess. When installed correctly, the ends of the insert stand equally proud above the parting face.

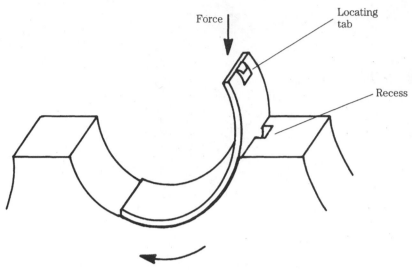

8-72 Main bearing inserts roll into place. Slip the bearing into the cap or saddle, leading with the smooth edge. Then, with thumbs providing the force, push the insert home, seating the locating tab in its recess.

6. Verify that the oil passage drilled in the web indexes with the oil port in the insert. A glance will suffice, but some shops go a step further and run a drill bit through the insert and into the web.

7. Install the remaining upper shells, one of which might be flanged for thrust. Check each insert for indicated size and, once they are installed, verify that oiling passages are open.

8. Where appropriate, install the upper half of the rear crankshaft seal, per manufacturer's instructions. Rope seals must be rolled into place, using a bar sized to match crankshaft journal diameter.

9. Saturate the bearings and seal with clean lube oil. (Some mechanics prefer to use a grease formulated for bearings assembly, such as Lubriplate No. 105. Grease is very appropriate when the engine will not be returned to immediate service.)

10. Wipe down the crankshaft with paper shop towels. Coat journals and crankpins with a generous amount of lubricant.

11. Lower the shaft gently and squarely into position on the webs. Exercise care not to damage thrust flanges.

12. Install lower bearing shells into the caps, lubricating as before.

13. Mount the caps in the correct orientation and sequence. Lightly oil cap bolt threads.

14. Torque the caps, working from the center cap outward. Conventional bolts make up in three steps—1/3, 2/3, and 3/3 torque; torque-to-yield bolts are run down to a prescribed torque limit and rotated past that limit by a set amount. After each cap is pulled down, turn the crankshaft to detect possible binds.

8-73 Lever the crankshaft through its full range of axial movement before measuring thrust bearing clearance. The same measurement can be made with dial indicator, registering off of the crankshaft nose. Chrysler Corp.

- *Radius ride,* a condition recognized by bright edges on the bearing shells and caused by excessively large crankshaft fillets left after regrinding. Return the crankshaft to the machinist.
- Dirt on the bearing face or OD.
- Insufficient running clearance, a condition that can be determined with Plasti-Gage (see below).
- Bent crankshaft or misaligned bearing saddles.

15. Using a pry bar, lever the crankshaft toward the front of the engine, then pull back (Figure 8-73). Measure crankshaft float with a feeler gauge between the thrust bearing face and crankshaft web.

Piston and rod installation

It is assumed that piston and rod assemblies have already been made up. If the machinist has not already done so, install the rings on the piston. Rings are packaged with detailed instructions, which supersede those in the factory manual. Here, it is enough to remind you that:

- The time required to check ring gap is well spent, because rings, like bearing inserts, are sometimes mislabeled. Using the flat of the piston as a pilot, insert each compression ring into the cylinder and compare ring gap with the specification (Figure 8-74).
- Ring orientation is important; according to one manufacturer, a reversed scraper ring increases oil consumption by 500%.
- Using the proper tool, expand the ring just enough to slip over the piston (Figure 8-75).
- Stagger ring gaps, as detailed by the manufacturer.

8-74 Measure ring gap relative to the lower, and least worn part of the cylinder. Yanmar Diesel Engine Co. Ltd.

8-75 Install piston rings in the sequence shown. Kohler of Kohler

At this point, you are ready to install the piston assemblies. Follow this procedure:

1. Remove the bearing cap from No. 1 rod and piston assembly
2. Mount an upper bearing shell on the rod, indexing oil ports.
3. Coat the entire bearing surface with fresh lube oil. Repeat the process for the cap.
4. Slip lengths of fuel line over the rod bolt ends to protect the crankshaft journals during installation.
5. Turn the crankshaft to bottom dead center on No. 1 crankpin. Saturate the crankpin with oil.
6. Install a ring compressor over the piston. The bottom edge of the compressor should be a ½ in. or so below the oil ring. Tighten the compressor bands just enough to overcome ring residual tension.
7. With the block upright and the leading side of the piston toward the front of the engine, place the compressor and captive piston over No. 1 cylinder bore. A helper should be stationed below to guide the rod end over the crankpin. While holding the compressor firmly against the fire deck, press the piston out of the tool and into the bore. As shown in Figure 8-76, thumb pressure should be sufficient. Stop if the piston binds and reposition the piston in the compressor.
8. Install the lower bearing shell in the cap; verify that the locating tab indexes with its groove.
9. Coat the bearing contact surface with lube oil.
10. Working from below, carefully remove the hoses from the bolt ends and pull the rod down over the crankpin. Make up the rod cap, making sure match-marks align.
11. Torque the rod bolts to spec (Figure 8-77).
12. Using a hammer handle or brass knocker, gently tap the sides of the big end journal. The rod should move easily along the length of the crankpin.

8-76 Piston installation should be a gentle process, involving no more than the force exerted by one hand. Peugeot

8-77 Using an accurate torque wrench, pull the rod bolts down evenly to spec. Yanmar Diesel Engine Co. Ltd.

13. Rotate the crankshaft a few revolutions to detect possible binds and the bore scratches that mean a broken ring. Resistance to turning will not be uniform: piston speed increases at mid-stroke, with a corresponding increase in crankshaft drag.
14. Repeat this operation for remaining pistons.
15. Press or bolt on piston cooling jets, aiming them as the factory manual indicates.

Plastigauge

Old-time mechanics checked timing gears by rolling a piece of solder between the gear teeth. The width of the solder represented gear lash. (This technique has generally been replaced by direct measurement with a dial indicator.) Plastic gauge wire represents an application of the same principle to journal bearings. Perfect Circle and several other manufacturers supply color-coded gauge wire together with the necessary scales that translate wire width into bearing clearance. Green wire responds to the normal clearance range of 0.0001–0.0003 in.; red goes somewhat higher—0.0002–0.0006; blue extends to 0.0009 in. Accuracy compares to that obtained with a micrometer.

1. Remove a bearing cap and wipe the lubricant off both the cap and the exposed portion of the journal.
2. Lay a strip of gauge wire longitudinally on the bearing, about ¼ in. off-center, as shown in Figure 8-78.

Plastic strip

8-78 Plastigauge wire is an inexpensive and quite accurate method of determining bearing clearance. Ditroit Diesel

Note: Invert the engine or raise the crankshaft with a jack under an adjacent counterweight before gauging main-bearing clearances. Otherwise, the weight of the crankshaft and flywheel will compress the wire and give false readings.

3. Assemble the cap, torquing the hold-down bolts to specifications.
4. Without turning the crankshaft, remove the cap. Using the scale printed on the gauge-wire envelope, read bearing clearance as a function of wire width. Clearance should fall within factory assembly specifications. If it does not, remove the crankshaft and return it to the machinist. Scrape or wipe off the gauge wire with a rag soaked in lacquer thinner. Reoil the bearing and journal before final assembly.

Balancing

Precision balancing is an optional service, difficult to justify in economic terms, but worth having done if the goal is to build the best possible engine. Balanced engines run smoother, should last longer, and sometimes exhibit a small gain in fuel economy.

But do not expect wonders; balancing is a palliative, arrived at by compromise, and effective over a fairly narrow rpm band. No amount of tweaking can take all of the shake, rattle, and roll out of recip engines.

There are three steps in the balancing process. First the technician equalizes piston and pin weights to within $1\frac{1}{2}$ g (Figure 8-79). If the technician works out of a large inventory, this can sometimes be achieved by selective assembly. Otherwise, machine work is in order; heavy pistons are lightened by turning the inner edge of the skirts, and when absolutely necessary, aluminum slugs can be pressed into the hollow piston pins to add weight to light assemblies.

8-79 Balancing begins with match-weighing piston assemblies.

The next step is to match-weight the rods using a precision scale and a special rod adapter, such as shown in Figure 8-80. The Stewart-Warner adapter makes it possible to distinguish between rotating (or lower-end) and reciprocating (upper-end) masses; when such equipment is not available, the technicians assume that the lower half of the rod rotates and that the upper end describes a purely reciprocating path. However arrived at, rod reciprocating and rotating masses are equalized to a 1½ g tolerance by removing metal from the balance pads (shown as the extensions on the ends of the DDA rod in the photograph and more clearly back in Figure 8-59, where they are labeled "noncritical" areas).

8-80 Stewart-Warner fixture splits rod weight into the rotating mass of piston and rod assemblies.

The technician computes the total rotating mass and bolts an equivalent mass to the crankshaft (Figure 8-81). The shaft is then mounted on the balancer and rotated at the desired speed, which should correspond to actual operating conditions. Balance, or matching counterweight mass to rotating mass, can normally be achieved by drilling the counterweights. Most shops work to a 0.5 oz.-in. tolerance; the equipment will support 0.2 oz.-in. accuracy. Once this is done, the process is repeated for the harmonic balancer and clutch assembly.

Oil seals

Oil seals can be a headache, especially if one leaks just after an overhaul. Premature failure is almost always the mechanic's fault and can be traced to improper installation.

8-81 The crankshaft is dynamically balanced relative to the rotating mass of piston and rod assemblies.

Rope or strip seals are still used at the aft end of the crankshaft on some engines. These seals depend on the resilience of the material for wiping action and so must be installed with the proper amount of compression. Remove the old seal from the grooves and install a new one with thumb pressure. (You might want to coat the groove—but not the seal face—with stickum.) The seal should not be twisted or locally bound in the groove.

Now comes the critical part. Obtain a mandrel of bearing boss diameter (journal plus twice the thickness of the inserts) and, with a soft-faced mallet, drive the seal home. Using this tool as a ram, you can cut the ends of the seal flush with the bearing parting line. Without one of these tools you will have to leave some of the seal protruding above the parting line in the elusive hope that the seal will compact as the caps are tightened. Undoubtedly some compaction does take place, but the seal ends turn down and become trapped between the bearing and cap, increasing the running clearance. Lubricate the seal with a grease containing molybdenum disulfide to assist in break-in.

Synthetic seals, usually made of Neoprene, are used on all full-circle applications and on the power takeoff end of many crankshafts as well. These seals work in a manner analogous to piston rings. They are preloaded to bear against the shaft and designed so that oil pressure on the wet side increases the force of contact.

These seals must be installed with the proper tools. If the seal must pass over a keyway, obtain a seal protector (a thin tube that slides over the shaft) or at least cover the keyway with masking tape. Drive the seal into place with a bar of the correct diameter. The numbered side is the driven side in most applications. The steep side of the lip profile is the wet side. It is good practice to use a nonhardening sealant on the back of metallic seals. Plastic coated seal housings are intended to conform to irregularities in the bore and do not need sealant.

8-82 Wear sleeves extend crankshaft life. When retrofitted, an oversized seal must be used to compensate for the increase in journal diameter. Detroit Diesel

More elaborate seals require special one-of-a-kind factory tools. The better engines often incorporate wear sleeves over the shafts, either as original equipment or as a field option. Figure 8-82 shows the use of a wear sleeve on a Detroit Diesel crankshaft. The sleeve makes an interference fit over the shaft and is further secured with a coating of shellac. Worn sleeves are cut off with a chisel or peened to stretch the metal. The latter method is preferred because there is less chance of damaging the shaft.

9
CHAPTER

Air systems

Diesel engines run lean, that is, with more air than required for chemically complete combustion. The surplus air passes through the engine and out the exhaust. But even the cleanest air contains mineral dust and other solid particles that, unless removed, cause rapid cylinder wear. These particles, or particulates as engineers call them, vary in size. The most destructive range between 10 and 20 μm in diameter. One micron equals one thousandth of a meter.

Air cleaners

The ideal air cleaner would trap all solids, regardless of size. Unfortunately, such a device does not exist, at least in practical form. Some particulates get through, which is why engines tend to wear rapidly in dusty environments, no matter how carefully maintained.

Oil-bath air cleaners were once almost universal, but now are pretty well confined to engines that operate in extremely dusty environments. These devices combine filtration with inertial separation (Figure. 9-1). Internal vanes in the precleaner impart rotation to the incoming air stream. Heavier particulates centrifuge out and collect in the fixed filter element. As the air stream approaches the bottom of the canister, it reverses direction. Many of the particulates that remain fail to make the U-turn and end in the oil reservoir. Those that are left must find their way through the oil-wetted replaceable filter element.

These devices depend, in great part, upon the velocity of the incoming air to centrifuge out heavier particles. At low engine speeds, the particles remain in the air stream, and filtration depends solely upon the oil-wetted mesh. Under ideal circumstances filtering efficiency is no more than 95% and, in practice, can be much less. If overfilled, subjected to high flow-rates, or tilted much off the horizontal, oil-bath

Centrifugal precleaner

Clip

Main body and fixed filter element

Gasket

Detachable filter element

Oil bath

9-1 Oil-bath air cleaner with a centrifugal precleaner. These units have a peak efficiency of about 95%, which falls off rapidly at low speeds.

cleaners bleed oil into the intake manifold. The oil can plug the aftercooler and raise exhaust temperatures enough to cause turbocharger failure.

Replaceable paper-element filters have become standard (Figure 9-2). If the designers have been thorough, the paper element includes a foam prefilter and may be combined with a centrifugal precleaner. At their most sophisticated, paper-element filters can have an efficiency of 99.99%. Efficiency improves when the filter is lightly impacted with dust. Engines operated in extremely dusty environments benefit from an exhaust-powered ejector mounted upstream of the filter. Some applications include a safety filter downstream of the main element. The safety filter acts as a backup in the event of damage to the primary filter and prevents dust entry when the main filter is serviced.

The air cleaner should mount horizontally on the manifold to reduce the possibility of dirt entry when the element is changed. Cleaners that mount vertically,

1 Cover assembly
2 Body packing
3 Element packing
4 Element assembly
5 Body assembly
6 Element packing
7 Rubber ring
8 Bolt
9 Rubber ring
10 Bolt
11 Lockwasher
12 Flatwasher

9-2 A single-stage, paper-element air cleaner of the type used for automotive and marine applications. Caterpillar, Inc.

such as the one shown in Figure 9-2, should include a semi-permanent safety element below the main filter. In addition, all engines should have some sort of filter monitoring device that, if not supplied by the manufacturer, can be fabricated by mounting a vacuum gauge immediately downstream of the filter. When new, paper-element filters impose a pressure drop of about 6.0 in./H_2O and the inlet ducting usually adds about 3.0 in./H_2O. Engine performance falls off noticeably when the total system pressure drop exceeds 15 in./H_2O.

Service the filter only when the air-restriction indicator trips, since each time the element is removed some dust enters the system. Before removing the element, wipe off any dust from the inside of the filter housing. Do not blow out paper elements with compressed air—a tear in the element, so small that it may not be visible, exposes the engine to massive amounts of dust intrusion.

And before we leave the subject, it should be remarked that "high-performance" filters provide marginal power increases by reducing pumping losses. But the reduction in pressure drop often comes at the cost of reduced filtration.

Any filter worthy of its name should have its efficiency certified by a third party under the protocols of the Society of Automotive Engineers (SAE) air cleaner test code for automotive dry and wet filters, and SAE J2554 for heavy-duty engines.

Flexible couplings should be replaced periodically and as inexpensive insurance for rebuilt engines (Figure 9-3). Secure the coupling with SAE Type F worm-gear stainless steel clamps—not plumber's clamps—that provide a 360° seal.

9-3 The air-inlet system should be constructed of large-diameter plastic tubing with fewest possible directional changes. Avoid tees and right-angle ells. Mounting air cleaners on the engine frame or chassis minimizes vibration damage, but if this is done, the inlet piping must be joined with flexible hose couplings. Caterpillar, Inc.

Air-inlet plumbing must have adequate capacity and a minimum of pressure drop. In the absence of factory data, engine flow requirements can be calculated as described by Caterpillar:

$$CFM = (CID^5 \times speed^2 \times VE^3 \times PF^4)/(1728 \times CF)$$

CFM = cubic ft/min
CID = cubic inch engine displacement
speed = maximum engine rpm
VE = volumetric efficiency
>2.0 = state-of-the-art engines
1.3 to 1.8 = typical turbocharged four-cycle
0.85 = naturally aspired four-cycle
1.4 = 2-cycle with Roots blower (e.g., vintage Detroit)
CF = cycle factor
2 = four-cycle
1 = four-cycle
PF = pulsation factor (three- or fewer cylinder naturally aspiraed four-cycle engines).

2.0–2.1 = single cylinder
1.0–1.4 = twin cylinder
   1.33 = three cylinders

Again following Caterpillar guidelines, the maximum permissible inlet restriction is 6 in./H_2O for on-highway engines and 10 in./H_2O for off-road or heavy industrial applications.

Turbochargers

A turbocharger is an exhaust-powered supercharger that, unlike conventional superchargers, has no mechanical connection to the engine (Figure 9-4). The exhaust stream, impinging against the turbine (or "hot") wheel, provides the energy to turn the compressor wheel. For reasons that have to do with the strength of materials, turbo

9-4 A turbocharger is an exhaust-driven centrifugal pump, operating at five- or six-figure speeds and typically generating 10 to 12 psi of boost.

boost is usually limited to 10 or 15 psi. This is enough to increase engine output by 30 or 40%. For example, the addition of a turbocharger to the John Deere 6068T naturally aspirated engine increased the output from 135 to 175 hp and torque from 335 lb-ft to 473 lb-ft.

Turbocharging is also a "green" technology, because the energy for air compression would otherwise be wasted as exhaust heat and noise. On the other hand, the interface between the engine and the turbo machinery, which turns at speeds on the order of 160,000 rpm (with the potential for 200,000-plus rpm) and experiences temperatures in excess of 1000°F, is not seamless.

The turbine wheel draws energy from exhaust gas velocity and heat, qualities that increase with piston speed and load. The compressor section behaves like other centrifugal pumps, in that pumping efficiency increases with impeller speed. At low speeds, a large fraction of the pumped air bleeds back through the clearance between the rim of the impeller and the housing. At very high rotational speeds, air takes on the characteristics of a viscous liquid and pumping efficiency approaches 100%. In its primitive form, a turbocharger acts like the apprentice helper, who loafs most of the day and, when things get busy, becomes overly enthusiastic.

Another characteristic of turbocharged engines is the lag, or flat spot, felt during snap acceleration. Perceptible time is required to overcome the inertia of the rotating masses of the turbine and compressor wheels. By the same token, the wheels continue to coastdown after the engine stops.

Background

The exhaust-driven supercharger was first demonstrated in 1915, but remained impractical until the late 1930s, when the U.S. Army, working closely with General Electric, developed a series of turbocharged aircraft engines. The expertise in metallurgy and high-temperature bearings gained in this project made GE a leader in turbocharging and contributed to its success with jet engines. As applied to aircraft, turbocharging was a kind of artificial lung that normalized manifold pressure at high altitudes. A wastegate deflected exhaust gases away from the turbine at low altitudes, where the dense air would send manifold pressures and engine power outputs to dangerously high levels. In emergencies, P-51 pilots were instructed to push their throttles full forward, an action that broke a restraining wire, closed the wastegate, richened the mixture, and initiated alcohol injection for a few seconds of turbo boost. The broken wire signalled the ground crew to rebuild the engine.

Turbochargers continue to be used on reciprocating aircraft engines. Currently about 20% of the cars and light trucks sold in the United States are turbocharged and more will be in the future as manufacturers downsize their vehicles to meet the 54.5 miles per gallon CAFE (Corporate Average Fuel Economy) standard due to come into effect in 2025 (Figure 9-5). According to Ward's Automotive 27.6% of cars and light trucks sold during the first-quarter of 2017 were turbocharged. Turbodiesels accounted for three-quarters of the number.

Turbocharging addresses the fundamental shortcoming of compression ignition, which is the delay between the onset of injection and ignition. That delay, terminated by the sudden explosion of puddled fuel, results in a rapid cylinder pressure rise, rough

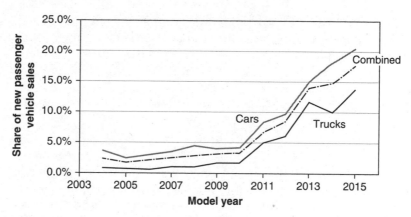

9-5 U.S. market share of turbocharge light-duty vehicles. EPA 2015 Fuel Economy Report.

running and incomplete combustion with attendant emissions. Turbocharging (and supercharging generally) increase the density of the air charge. Ignition temperature develops early during the compression stroke and coincides more closely with the onset of injection. Fuel burns almost as quickly as delivered for a smoother running and more economical engine.

Depending on how it is accomplished, turbocharging can have three quite distinct effects on performance. If no or little additional fuel is supplied, power output remains static, but emissions go down. Surplus air lowers combustion temperatures and provides internal cooling. Such engines should be more durable than their naturally aspirated equivalents. This high-boost/low-fuel approach to turbocharging is limited to giant stationary and marine plants.

Makers of mid-sized, high-speed engines have been more concerned with maximum power, mid-range torque or emissions control. Supplying additional fuel in proportion to boost yields power, which can translate as fuel savings for engines that run under constant load. Some gains in fuel efficiency (calculated on an hp/hour basis) accrue from turbocharging, but the real advantage comes about when high supercharge pressures allow for lower piston speeds. Large marine engines, developing a 1000 hp and more per cylinder, use this approach to achieve thermal efficiencies of 50%. On a more familiar scale, the naturally aspirated International DT-414 truck engine produced 157 hp at 3000 rpm. The addition of a large turbocharger boosted output to 220 hp, for a gain of 40%. Once in the fairly narrow power band, the trucker could save fuel by selecting a higher gear.

The third approach is characteristic of automotive and light truck engines, which operate under varying loads and rarely, if ever, develop full rated power. What is wanted for these applications is torque.

Torque is the quality that makes driving fun. Mid-range torque was the quality that propelled diesel passenger cars into first place in European sales and enabled them to establish a beach head in the United States where emissions rules and fuel prices were less favorable.

The Navistar 7.3L, developed for Ford pickups and Econoline vans, is perhaps the best demonstration of how a turbocharger can be throttled for torque production. In its naturally aspirated form, the engine develops 185 hp at 3000 rpm and 360 ft/lb. of torque at 1400 rpm. A Garrett TC43 turbocharger increases power output marginally to 190 hp; but torque goes up 17.8% to 385 ft/lb. In normal operation the wastegate remains closed, directing exhaust gases to the turbine, until 1400 rpm. At that point, which corresponds to the torque peak, the hydraulically actuated wastegate begins to open, shunting exhaust away from the turbine.

Boosts in power or torque impose severe mechanical and thermal loads, which should be anticipated in the design stage. Add-on turbocharger kits—which rarely involve more than a rearrangement of the plumbing—are a buyer-beware proposition. The 7.3L is considered a very rugged engine in its naturally aspirated form. But turbocharging called for hundreds of engineering changes, including shot-peened connecting rods, oversized piston pins, Inconel exhaust valves, and special Zollner pistons with anodized crowns.

Operation

Figure 9-6 illustrates an air-cooled Ishikawajima-Harima turbocharger of the type found on engines in the 100–150-hp range. The inset shows the water-cooled version of the same turbocharger, plumbed into the engine cooling system.

Compressor   The compressor section consists of an impellor backed by a vaned diffuser and encased in a spiral housing, or volute. The rapidly spinning hub draws air into the impellor hub where it spills out to into channels formed by the impellor blades. Diffuser blades both slow the airflow velocity and direct the airflow to the volute, where it undergoes further deceleration. The loss of velocity creates pressure and heat.

The ratio between turbo inlet air pressure and outlet pressure is the pressure ratio. A low airflow rate and a relatively high pressure ratio result in *surge*. The compressed air momentarily reverses the direction of the incoming air stream. Surge is the low-rpm limit of compressor performance and typically occurs when the turbo is too large for the application.

The high-rpm compressor limit occurs when the large volume of airflow becomes more than the turbo can pass. Unable to breathe, the turbo experiences *choke*. A turbo announces that it is too small for the job by choking at wide open throttle.

Nearly all compressor wheels are consist of C355 or 354 HIP aluminum alloy investment castings. For limited production runs manufacturers resort to billet wheels machined from 2618 aluminum forgings (Figure 9-7). Titanium impellors are specified for extreme service. While Ti is denser than Al, thinner vanes compensate so that there is little difference in spool-up time.

Turbine   The rule of thumb that exhaust gas temperature (EGT) should not exceed 1250°F (667°C) is rapidly becoming obsolete. When stressed to the maximum, Cummins/Dodge truck EGTs can register exceed 1300°F on a pyrometer installed upstream of the turbo. Turbine wheels must live for hundreds of hours in these temperatures while resisting the corrosive effects of exhaust gas and the centrifugal force induced by spinning at 150,000 rpm and higher speeds.

Clamp (compressor side)

M6, bolt

Blower wheel

Oil thrower

Thrust busing

Lock nut

M4, screw

Blower wheel chamber

Retaining ring

Seal plate

Bearing chamber

Seal ring(small)

Seal ring(large)

Heat insulating board

M3, screw

Thrust bearing

Seal ring (turbine side)

Floating bearing

Turbine shaft

Turbine chamber

M6, bolt

Clamp (turbine side)

Cooling water

Charging air

Exhaust gas

9-6 Air- or water-cooled turbocharger of the type used on small Japanese diesels. Water cooling should prolong bearing life and reduce the coking experienced during hot shutdowns.

9-7 A Honeywell titanium compressor wheel of the type used for medium-duty engines. Ti is denser than aluminum, but creeps less with heat, has better fatigue resistance and is less susceptible to the corrosion induced when exhaust gas is recirculated through the compressor.

Hot wheels are vacuum cast from high-strength nickel-chrome-based alloys, such as Inconel 713C or 713LC. Machining is limited to minor cleanup operations and to drilling the hub hole, which centers on the wheel mass for balance. Cummins joins the turbine wheel and shaft by friction welding, a process that produces a stronger joint than the parent metals. The new Borg Warner EFR turbo breaks tradition with a titanium-aluminide turbine wheel, said to weigh half as much as the equivalent Inconel part.

In order to survive the heat and, should the worst happen, contain the fragments, turbine housings are made of cast iron or investment-cast stainless steel. Stainless castings weigh a third less than iron and retain heat better, but cost considerably more.

Bearing section Most turbos have floating sleeve bearings with friction surfaces on both the ID (internal diameter) and OD (external diameter). Floating bearings reduce rubbing speed by half and, interestingly enough, were used on first-generation Ford V-8 connecting rods. Annular-contact ball bearings found on better turbochargers generate less friction than bushings and are better able to tolerate the lack of oil pressure during startup. Garrett and other manufacturers now supply ceramic ball bearings that, because of their lighter mass, are less likely lift under centrifugal force and lose rolling contact with their inner races.[1]

As shown in Figure 9-8, most turbos employ piston-ring seals, although gas-lubricated face seals are an option.

Wastegates Fixed geometry turbochargers usually have a wastegate that shunts exhaust gas around the turbine to prevent excessive boost. The wastegate may be integral with the turbine housing (Figure 9-9) or mounted on dedicated intake-manifold runner. The latter configuration permits a more compact turbine housing for rapid spool-up.

Another way to limit boost is to bleed some of air off the compressor with a groove or a series of holes in the housing that convey air back to the inlet (Figure 9-10).

Compressor End Seal System   **Turbine End Seal System**

9-8 Turbo shaft seals, intended to primarily to keep exhaust gas and boost air out of the crankcase and secondarily to prevent oil intrusion into the intake manifold, function like piston rings.

---

[1]Kane, J., "Turbochargers," EPI, May 5, 2017, http://www.epi-eng.com/piston_engine_technology/turbocharger_technology.htm

Relief valve

Exhaust gas to turbine

Compressor wheel

Oil

Turbine wheel

Diaphragm

Intake air

Wastegate valve

Wastegate actuator

Exhaust gas from engine

9-9 Diaphragm-controlled wastegate and pressure-relief valve, used in on Ford 2.3L diesel engines. A small hose transfers boost pressure from the compressor outlet to a spring-loaded diaphragm in the wastegate actuator. When boost pressure exceeds spring force, the wastegate opens. A disadvantage of this arrangement is that the wastegate unseats to bleed off exhaust before the specified boost pressure is reached. An actuator designed to fully open the wastegate at 10-psi may crack it open at half that value. More sophisticated actuators actuate by a steeper motor or solenoid under command of the ECU (electronic control module).

Normal air flow

Re-circulated airflow for surge control

9-10 A Garrett turbocharger with a ported shroud.

While short-circuiting, the compressor has an adverse effect upon efficiency and can result in noisy operation, many manufacturers resort to it.

Compressor map

The complex interrelationships between pressure ratio, air flow, turbo speed, and compressor efficiency can be visualized on a "map" such as the one shown in Figure 9-11. The pressure ratio shown on the vertical scale is outlet pressure divided

9-11 A sample turbo map.

by inlet pressure. Note that these pressures are absolute with zero representing a pure vacuum. Boost gauges reference zero to atmospheric air pressure, or 14.7 psi at sea level. A boost pressure of 10 psi at sea level means that absolute manifold pressure is 14.7 psi + 10 psi, or 24.7 psi. The turbo pressure ratio is then:

$$24.7 \text{ output} \div 14.7 \text{ intake} = 1.68$$

The horizontal axis is air flow corrected for density and the restriction imposed by the air filter. Surge and choke lines define the turbo operating envelope, and efficiency islands graph impellor efficiencies.

VGTs

Imagine for a moment that the exhaust system is a garden hose and the turbine inlet is the hose nozzle. Variable geometry turbochargers (VGTs) have moveable vanes that swivel or slide closed to reduce the area of the turbine inlet as the engine control module (ECM) commands. As with a water hose, a restriction reduces the flow volume and increases its velocity. Less water or exhaust gas gets through, but what does has more kinetic energy. And pressure upstream of the restriction—exhaust backpressure—increases. Opening the turbine inlet wider increases the flow rate and decreases flow velocity and backpressure.

Engineers describe VTGs as having variable A/Rs. As shown in Figure 9-12, the A/R is the ratio of turbine wheel radius to the cross-sectional area of the turbine inlet. The larger the ratio, the greater will be the area of the inlet relative to the turbine wheel.

VGT inlet vanes open high engine speeds to provide a large A/R ratio for maximum exhaust flow and boost (Figure 9-13A). The greater volume of exhaust gas available at high rpm compensates for the loss of gas velocity. At part throttle, the adjuster ring (4) swivels or slides the vanes (3) closed to reduce the A/R ratio (Figure 9-13B). To revert to our analogy, the garden hose has been choked down. The effectively

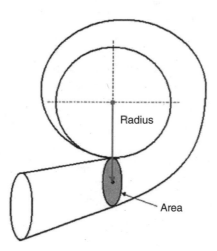

9-12 Changes in the A/R ratio trade off exhaust flow rate against flow velocity and backpressure. Honeywell Turbo Technologies.

9-13 A Bosch VGT in the open, or full boost (A) and closed (B) positions.

smaller turbocharger amplifies low- and mid-range torque for better acceleration and reduced turbo lag. Closing the vanes also creates backpressure in the exhaust system that augments engine braking and makes exhaust gas recirculation (EGR) possible.

Recycling a portion of the exhaust gas into the combustion chambers is the primary means of controlling nitrogen oxide emissions. Early on, EGR was resorted to only at low speeds, when small amounts of intake manifold vacuum are present. All that was necessary was a metered connection between the exhaust system and the intake manifold. But 2004 EPA Tier 2 regulations mandated EGR at higher engine speeds where there is little or no manifold vacuum. The VGT, with its ability to produce backpressure at will, came to the rescue.

Vane pitch is controlled with a stepper motor or by lube oil or compressed air acting on a spring-loaded diaphragm. A rod transfers diaphragm movement to the vane control ring. Replacement actuators include instructions for adjustment that, in some cases, can require a scanner capable of cycling the unit and a precisely regulated source of pressure. Figure 9-14 shows a Detroit Diesel air-operated VGT actuator. How this actuator is controlled is illustrated in Figure 9-15.

9-14 Detroit Diesel air-operated VGT actuator.

9-15 This diagram illustrates VGT control inputs for particulate matter (PM) and oxides of nitrogen (NO_x) reduction. The controller includes look up tables (LU) and a microprocessor unit (MPU) with random-access memory (RAM), non-volatile random access memory (NVRAM), and read-only memory (ROM). The controller receives inputs from various engine sensors, which are not shown in the drawing. The MAP sensor reports intake-manifold air flow (AF), the backpressure sensor reports backpressure (BP), the EGR valve reports its position (POSIT), the VGT turbo its vane position (VAPOS), and the DPF, or diesel particulate filter, its pressure drop (DP). The controller integrates this data in output commands adjust the variable orifice in the EGR valve (ACT) and the VGT vane position (ADJ).

VGTs work beautifully when they work. But problems occur with close-toleranced parts in a hostile environment. Vanes stick opens with soot or rust due to the corrosive effects of exhaust gas and condensation. A thorough scrubbing with EGR cleaner and mild Scotch Brite scouring pads is usually enough to put the turbocharger back into service. Another common complaint is excessive play in at the actuator arm/control plate connection. The arm attaches to a hole in the vane ring that, after long use, elongates enough to limit vane movement.

Heat-damaged actuators must be replaced, a procedure that involves setting the actuator-arm length to specification. Normally this can be done with vernier calipers, although in the case of the Holset HX55V space limitations require the use of a special tool.

Aftercoolers

Compressing air raises its temperature, which reduces change density and tends to defeat the purpose of supercharging. Most turbocharger installations include a heat exchanger, or aftercooler, between the compressor outlet and intake manifold. The cooling medium can be air, engine coolant, or—for marine applications—water. The plate and tube seawater-fed aftercooler used with the Yanmar 4LH-HTE boosts output to 135 hp, or 30 hp more than an identical engine without aftercooling.

Troubleshooting

Table 9-1 lists the symptoms of commonly encountered turbocharger faults.

Turbo support system inspection

Lubrication system   A turbocharger makes severe demands on the lubricating system. Elevated combustion pressure contaminates the oil with blowby gases, and promotes oxidation by raising crankcase oil temperature. That fraction of the oil diverted to the turbocharger can undergo 80°F temperature rise in its passage over the shaft bushings. AHPD (super-high-performance-diesel) oils are superior to the API-CD oils. Anecdotal evidence suggests that Mobile 1 5W-40 Turbo Truck Diesel can pay for itself with longer oil change intervals.

Maximum allowable oil temperature varies with the application, but 250°F is the generally agreed-upon limit. Turbo bearings need oil within 4 seconds of startup and survive best with normal oil pressures of 30 psi or higher. Idle oil pressure should be at least 15 psi. Leaks, kinks, or partial stoppages in the oil supply and oil return lines are fatal. Oil-return plumbing should be essentially vertical (the closer to vertical the better) and terminate at a point well above the crankcase oil level.

Table 9-1. Turbocharger fault diagnosis

Engine lacks power, black smoke in exhaust

Symptom	Probable causes	Corrective actions
1. Insufficient boost pressure; compressor wheel coasts to a smooth stop when engine is shut down; turns easily by hand.	1. Clogged air filter element.	Clean or replace element.
	2. Restriction in air intake.	Remove restriction.
	3. Air leak downstream of compressor.	Repair leak.
	4. Insufficient exhaust gas energy	
	• Exhaust leaks upstream of turbocharger.	Repair leak.
	• Exhaust restriction downstream of turbocharger.	Remove restriction.
2. Insufficient boost pressure; compressor wheel does not turn or judders during coast-down; drags or binds when turned by hand.	1. Carbon accumulations on turbine-shaft oil seals.	Disassemble, clean turbocharger and change oil and filter.
	2. Bearing failure, traceable to	
	• Normal wear.	Rebuild turbo.
	• Insufficient oil.	Rebuild turbo. Inspect oil supply/return circuit and repair as necessary to restore full flow. Change engine oil and filter.
	• Excessive oil temp.	Rebuild turbo. Clean oil cooler (if fitted); change engine oil and filter. Monitor oil temp.
	• Turbo rotating assy out of balance.	Rebuild and balance turbo.
	• Bad operating practices— application of full throttle upon startup and/or hot shutdown.	Rebuild turbo. Train operators. A prelube system may extend turbo life in harsh operating environments.
	3. Rotating assembly makes rubbing contact with case. Leading causes are	
	• Bearing failure (see above).	Rebuild turbo. Determine cause of failure and correct.
	• Entry of foreign matter into compressor.	Rebuild turbo. Inspect air cleaner and repair as necessary.
	• Excessive turbo speed.	Rebuild turbo. Turbo rpm is a function of exhaust gas temperature (engine load). Check air filter for restrictions, bar over engine to verify that crankshaft rotates freely. If necessary, adjust governor to reduce engine power output.
	• Improper assembly.	Repair turbo, review-service procedures.

(Continued)

Table 9-1. Turbocharger fault diagnosis (*Continued*)

Excessive oil consumption, blue or white exhaust smoke

Symptom	Probable causes	Corrective actions
1. Condition may be accompanied by oil stains in the inlet and/or outlet ducting, or, in extreme eases, by oil drips from the turbo- charger housing.	1. High restriction in the air inlet. 2. Worn turbocharger seals, possibly associated with bearing failure and/or wheel imbalance.	Clean or repalce filter element. Repair turbo.

Abnormal turbocharger lag

Symptom	Probable cause	Corrective actions
1. Engine power output exhibits a pronounced "flat spot" during acceleration. Fuel system appears to function normally.	1. Carbon buildup on compressor wheel and housing.	Disassemble and clean turbo.

Unusual noise or vibration

Symptom	Probable causes	Corrective actions
1. "Chuffing" noise, often most pronounced during acceleration.	1. Surging caused by restriction at compressor discharge nozzle.	Disassemble and clean turbo.
2. Knocking or squeal.	1. Rotating elements in contact with housing because of bearing failure or impact damage.	Replace or rebuild turbo.
3. Speed-sensitive vibration.	1. Loose turbocharger mounts 2. Rotating elements in contact with housing. 3. Severe turbine shaft imbalance.	Tighten Replace or rebuild turbo. Rebuild and balance.

It is good practice to clean oil supply and return lines every 35,000 miles or even more frequently. Restrictions in the oil supply can result in bearing failure at full boost. Should this happen, the shaft binds and the terrific inertia of the spinning wheels snaps the turbo shaft in two.

Excessive oil consumption accompanied by oil puddling in the compressor section and by smoke in the exhaust may be caused by turbo seal failure. But a more likely cause is cylinder wear that overwhelms the PVC system with blowby. Make a cylinder leakdown test or a compression test before condemning the turbo seals. Another possible cause of oil burning is sludge in the turbo oil-return line.

The good news is that turbo piston-ring seals can leak without doing permanent damage to themselves. Fix the cause of the leak, and the seals will usually be okay.

When turbo seals do fail, expect to find loose bearings or carbon-cut grooves in the seal faces. Carbon damage results from hot shutdowns.

PCV The positive crankcase ventilation system removes combustion residues and, in the process, subjects the crankcase to a slight vacuum. In normal operation, fresh air enters through the breather filter and crankcase vapors discharge to the intake side of the turbo compressor. Excessive blowby due to leakage around the cylinder piston rings can flood the turbocharger with oil.

PCV system malfunctions can also involve the breather filter. Under severe load, blowby gases may accumulate faster than they can be vented and discharge back through the breather filter. These flow reversals, which occur more frequently in turbocharged engines, clog the filter. The resulting high crankcase pressure forces oil past the crankshaft seals and crankcase gaskets.

Air inlet system Dust particles, entering through a poorly maintained filter or through leaks in the ducting, rapidly erode the compressor wheel. Use the factory-recommended air filter. "High performance" low-restriction filters should be viewed with skepticism since the easiest way to reduce the pressure drop across the filter is to make the element more porous. Repair air leaks without the use of gasket sealant, which can attack the oil seals.

Exhaust system Restrictions downstream of the turbine—crimped pipes, abrupt changes in direction, clogged mufflers—reduce turbo efficiency and, in extreme cases, represent enough load to reduce boost. However, exhaust leaks between the engine and turbocharger are more typical. The pipe rusts out (especially if wrapped in insulation), fatigues, or cracks under the 1000°F-plus temperature. If original equipment manufacturer (OEM) parts do not give satisfactory service, you might try calling a Flexonics applications engineer (800-533-1024). Exhaust tubing and bellows-type expansion joints are among the company's products.

Preliminary turbo inspection

If you feel it is necessary to observe compressor rotation, cover the turbocharger inlet with a screen to keep fingers and other solid objects out of the mechanism. When dismantling a turbocharger and related hardware, make a careful tally of all fasteners, lockwashers, and small parts removed. Be absolutely certain that all are accounted for before starting the engine. Immediately shut down the engine if the turbocharger makes unusual noise or vibrates.

The seven-step inspection procedure outlined here was adapted, with modifications, from material supplied by John Deere:

1. *Turbo housing* Before disconnecting the oil lines, examine external surfaces of the housing for oil leaks, which would almost certainly mean a turbo seal or related failure.

2. *Compressor housing inlet and wheel*   Inspect the compressor wheel for erosion and impact damage. Erosion comes about because of dust intrusion; impact damage is prima facie evidence of negligence. Carefully examine the housing ID and compressor blade tips for evidence of rubbing contact, which is the definitive sign of bearing failure.

3. *Compressor housing outlet*   Check the compressor outlet for dirt, oil, and carbon accumulations. Dirt points to a filtration failure; oil suggests seal failure, although other possibilities exist, such as clogged turbo-oil return line or crankcase breather. Carbon on the compressor wheel might suggest some sort of combustion abnormality, but the phenomenon is also seen on healthy engines.

4. *Turbine housing inlet*   Inspect the inlet ports for oil, heavy carbon deposits, and erosion. Any of these symptoms suggests an engine malfunction.

5. *Turbine housing outlet and wheel*   Examine the blades for impact damage. Look for evidence of rubbing contact between the turbine wheel and the housing, which would indicate bearing failure.

6. *Oil-return port*   The shaft is visible on most turbochargers from the oil-return port. Excessive bluing or coking suggests lubrication failure, quite possibly associated with hot shutdowns.

7. *Bearing play measurements*   Experienced mechanics determine bearing condition by feel without dismantling the turbo. Use of a dial indicator gives more reliable results. Measurement of radial, or side-to-side, bearing clearance involves moving the shaft from one travel extreme to another, 180° away (Figure 9-16A). Hold the shaft level during this operation: a rocking motion would muddy the results. Axial motion is measured as travel between shaft thrust faces (Figure 9-16B). In very general terms, subject to correction by factory data for the unit in question, we would be comfortable with 0.002 in. radial and 0.001 in. axial play. Schwitzer, Garrett, and small foreign turbos tend to be set up tighter.

Overhaul

General instructions are almost meaningless since construction details, wear limits, and torque specifications vary so much between make and model. Most Holset manuals are available free for downloading on the Internet, as are some Schwitzer and Garrett manuals.

What does apply generally is that turbo housings and compressor wheel/shaft relationships must be marked prior to disassembly. The central nut that secures the compressor wheel to the shaft usually has a left-hand thread. Loosen and tighten the nut with a double-jointed socket-wrench extension to isolate the shaft from bending forces. Clean the parts with solvent and a plastic scraper. Metal scrapers or wire wheels must not be used.

WARNING: Wear eye protection when dealing with snap rings.

9-16 A dial indicator is used to measure radial (A) and axial (B) shaft play, specified as total indicator movement. Take off the radial measurement at a point near the center of the shaft, with shaft held level throughout its range of travel. Yanmar Diesel Engines Co., Ltd.

    The most critical part is the compressor-side oil seal and seal plate. A massive oil leak across this seal can result in a runaway. Feeding from crankcase oil, the engine accelerates ever faster until the bearings seize or a connecting rod breaks. A vehicle engine can sometimes be stopped by putting the manual transmission in the highest gear, holding down the brakes hard, and releasing the clutch. Stuffing rags into the air inlet or flooding the manifold with CO_2 from a fire extinguisher may also work. But hand-feeding an engine that verges on grenading is not the surest way of surviving into old age. It might be better to stand clear and let nature take its course.

9-17 A Honeywell ball-bearing CHRA.

The ready availability of center housing rotating assemblies (CHRAs) has simplified turbocharger repair. In the past, one had to rebuild from scratch, a piece at a time with opportunities for error at each step. CHRAs consist of a pre-assembled and balanced compressor wheel, shaft, bearing housing and turbine wheel (Figure 9-17). These units are available for popular BorgWarner, Honeywell Garrett, IHI (including the 2001-3 Duramax), MHI, and Schwitzer. OEM parts prices average around $600 and peak at $1000-plus for certain BorgWarner applications. Avoid off-brand "bargains."

10
CHAPTER

Electrical fundamentals

Diesel engines that develop more than 10 hp or so have a battery-operated starter and a generator to recharge the battery. The circuit may include glow plugs for cold starting and some basic instrumentation. The more sophisticated engines verge on robotic, in that one or more onboard computers determine fuel delivery and timing, turbo boost, and other critical parameters.

Merely looking at a starter motor or a heat sensor will tell you precious little about how it works. The only way to become even remotely competent in electrical work is to have some knowledge of the theory, of the interrelationship between volts, amperes and resistance and between electrical current and magnetism. This chapter is a brief, almost entirely nonmathematical, attempt to provide that information.

Electrons

Atoms are the building blocks of all matter. These atoms are widely distributed; if we enlarged the scale to make atoms the size of pinheads, there would be approximately one atom per cubic yard of nothingness. But as tiny and as few as they are, atoms (or molecules, which are atoms in combination) are responsible for the characteristics of matter. The density of a substance, its chemical stability, thermal and electrical conductivity, color, hardness, and all its other characteristics are fixed by the atomic structure.

The atom is composed of numerous subatomic particles. Using high-energy disintegration techniques, scientists are discovering new particles almost on an annual basis. Some are reverse images of the others; some exist for only a few millionths of a second. But, we are only interested in the relatively gross particles whose behavior has been reasonably well understood for generations.

In broad terms the atom consists of a nucleus and one or more electrons in orbit around it. The nucleus has at least one positively charged *proton* and might have one or more electrically neutral *neutrons*. These particles make up most of the atom's mass. The orbiting *electrons* have a negative charge. All electrons, as far as we

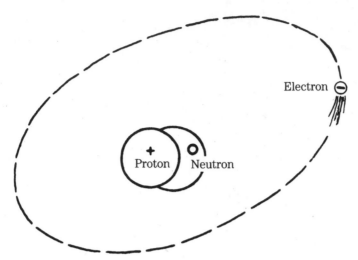

10-1 Representation of an electron orbiting around a nucleus consisting of a proton and a neutron.

know, are identical. All have the same electrical potency. Their orbits are balanced by centripetal force and the pull of the positive charge of the nucleus (Figure 10-1).

A fundamental electrical law states unlike charges attract, like charges repel. Thus two electrons, both carrying negative charges, repel each other. Attraction between opposites and the repulsion of likes are the forces that drive electrons through circuits.

Circuits

The term *circuit* comes from a Latin root meaning *circle*. It is descriptive because the path of electrons through a conductor is always in the form of a closed loop, bringing them back where they started.

A battery or generator has two *terminals,* or posts. One terminal is negative. It has a surplus of electrons available. The other terminal has a relative scarcity of electrons and is positive.

A *circuit* is a pathway between negative and positive terminals. The relative ease with which the electrons move along the circuit is determined by several factors. One of the most important is the nature of the conductor. Some materials are better conductors than others. Silver is at the top of the list, followed by copper, aluminum, iron, and lead. These and other conductors stand in contrast to that class of materials known as insulators. Most gases, including air, wood, rubber, and other organic materials, most plastics, mica, and slate are insulators of varying effectiveness.

Another class of materials is known as *semiconductors*. They share the characteristics of both conductors and insulators, and under the right conditions can act as either. Silicon and germanium semiconductors are used to convert alternating current (AC) to direct current (DC) at the alternator and to limit current and voltage outputs. These devices are discussed in more detail later in this chapter.

The ability of materials to pass electrons depends on their atomic structures. Electrons flood into the circuit from the negative pole. They encounter other electrons already in orbit. The newcomers displace the orbiting electrons (like repels like) toward the positive terminal and, in turn, are captured by the positive charge on the nucleus (unlike attracts). This game of musical chairs continues until the circuit is broken or until the voltages on the terminals equalize. Gold, copper, and other conductors give up their captive electrons with a minimum of fuss. Insulators hold their electrons tightly in orbit and resist the flow of current.

You should not have the impression that electrons flow through various substances in a go/no-go manner. All known materials have some reluctance to give up their orbiting electrons. This reluctance is lessened by extreme cold, but never reaches zero. And all insulators "leak" to some degree. A few vagrant electrons will pass through the heaviest, most inert insulation. In most cases the leakage is too small to be significant. But it exists and can be accelerated by moisture saturation and by chemical changes in the insulator.

Circuit characteristics

Circuits, from the simplest to the most convoluted, share these characteristics:
- Electrons move from the negative to the positive terminal. Actually the direction of electron movement makes little difference. In fact, for many years it was thought that electrons moved in the opposite direction—from positive to negative.
- The circuit must be complete and unbroken for electrons to flow. An incomplete circuit is described as *open*. The circuit might be opened deliberately by the action of a switch or a fuse, or it might open of its own accord as in the case of a loose connection or a broken wire. Because circuits must have some rationale besides the movement of electrons from one pole to the other, they have *loads*, which convert electron movement into heat, light, magnetic flux, or some other useful quantity. These working elements of the circuit are also known as *sinks*. They absorb, or sink, energy and are distinguished from the sources (battery and the generator). A circuit in which the load is bypassed is described as *shorted*. Electrons take the easiest, and least resistive path to the positive terminal.
- Circuits can feed single or multiple loads. The arrangement of the loads determines the circuit type.

Series and parallel circuits

A series circuit has its loads connected one after the other like beads on a string (Figure 10-2A). There is only one path for the electrons between the negative and positive terminals. In a string of old-fashioned Christmas tree lights, each lamp is rated at 10V; 11 in. series required 110V. If any lamp failed, the string went dark.

Such pure series circuits are rare today. (They can still be found in high-voltage applications such as airport runway lights.) But switches, rheostats, relays, fuses, and other circuit controls are necessarily in series with the loads they control.

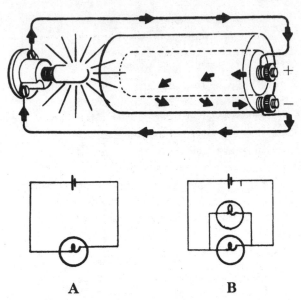

A   **B**

10-2 Basic series circuit (top) and its schematic (A). In B, a second lamp has been added with the first.

Parallel circuits have the loads arranged like the rungs of a ladder, to provide multiple paths for current (Figure 10-2). When loads are connected in parallel, we say they are *shunt*. Circuits associated with diesel engines usually consist of parallel loads and series control elements. The major advantage of the parallel arrangement from a serviceman's point of view is that a single load can open without affecting the other loads. A second advantage is that the voltage remains constant throughout the network.

Single- and two-wire circuits

However the circuit is arranged—series, parallel, or in some combination of both—it must form a complete path between the positive and negative terminals of the source. This requirement does not mean the conductor must be composed entirely of electrical wire. The engine block, transmission, and mounting frame are not the best conductors from the point of view of their atomic structures. But because of their vast cross-sectional area, these components have almost zero resistance.

In the single-wire system the battery is *grounded*, or as the British say, *earthed* (Figure 10-3). These terms seem to have originated from the power station practice of using the earth as a return. Modern engines have their negative battery post grounded. The "hot" cable connects the positive post to the individual loads, which are grounded. Electrons flow from the negative terminal, through the loads, and back to the battery via the wiring.

The single-wire approach combines the virtues of simplicity and economy. But it has drawbacks. Perhaps the most consequential is the tendency for the connections to develop high resistances. One would think that a heavy strap bolted to the engine

10-3 The principle of grounding.

block or frame would offer little resistance. But the connection is critical: a loose bolt or thin film of oil or rust on the mating surfaces can impede the flow of current.

Corrosion problems are made more serious by weather exposure and *electrolysis.* Electrolysis is the same phenomenon as occurs during electroplating. When current passes through two dissimilar metals (e.g., copper and cast iron) in the presence of damp air, one of the metals tends to disintegrate. In the process it undergoes chemical changes that makes it a very poor conductor.

Another disadvantage of the single-wire system, and the reason it is not used indiscriminately on aircraft and ocean-going vessels, is the danger of short circuits. Contact between an uninsulated hot wire and the ground will shunt the loads down circuit.

The *two-wire system* (one wire to the load, and a second wire from it to the positive terminal) is preferred for critical loads and can be used in conjunction with a grounded system. Electrolytic action is almost nil, and shorts occur only in the unlikely event that two bare wires touch.

Electrical measurements

Voltage is a measure of electrical pressure. In many respects, it is analogous to hydraulic pressure. The unit of voltage is the volt, abbreviated V. Thus we speak of a 12V or 24V system. Prefixes, keyed to the decimal system, expand the terms so we do not need to contend with a series of zeros. The prefix *k* stands for *kilo,* or 1000. A 50kV powerline delivers 50,000V. At the other end of the scale, *milli* means $^1/_{1000}$ and is abbreviated *m*. One millivolt (1 mV) is a thousandth of a volt.

The *ampere,* shortened sometimes to *amp* and abbreviated *A* is a measure of the quantity of electrons flowing past a given point in the circuit per second. One ampere represents the flow of 6.25×10^{18} electrons per second. Amperage is also referred to as *current intensity* or *quantity*. From the point of view of the loads on the circuit, the amperage is the *draw*. A free-running starter motor might draw 100A, and four or five times as much under cranking loads.

Resistance is measured in units named after G. S. Ohm. Ohms are expressed by the last letter of the Greek alphabet, omega (Ω). Thus we might speak of a 200Ω resistance.

The resistance of a circuit determines the amount of current that flows for a given applied voltage. The resistance depends on the atomic structure of the conductor—how tightly the electrons are held captive in their orbits—and on certain physical characteristics. The broader the cross-sectional area of the conductor, the lesser opposition to the current. And the longer the path formed by the circuit between the poles of the voltage source, the more is the resistance. Think of these two dimensions in terms of ordinary plumbing. The *resistance* to the flow of a liquid in a pipe is inversely related to its diameter (decreases as diameter increases) and directly related to its length. Resistance in the pipe produces heat, exactly as does resistance in an electrical conductor.

Resistance is generally thought of as the electrical equivalent of friction—a kind of excise tax that we must pay to have electron movement. There are, however, positive uses of resistance. Resistive elements can be deliberately introduced in the circuit to reduce current in order to protect delicate components. The heating effect of resistance is used in soldering guns and irons and in the glow plugs employed as starting aids in diesel engines.

Ohm's law

The fundamental law of simple circuits was expressed by the Frenchman G. S. Ohm in the early 1800s in a paper on the effects of heat on resistance. The law takes three algebraic forms, each based on the following relationship—a potential of 1V drives 1A through a resistance of 1Ω. In the equations, the symbol E stands for *electromotive force*, or, as we now say, volts. I represents *intensity*, or current, and R is resistance in ohms.

The basic relationship is expressed as:

$$I = \frac{E}{R}$$

This form of the equation states that current in amperes equals voltage in volts divided by resistance in ohms. If a circuit with a resistance of 6Ω is connected across a 12V source, the current is 2A (12/6 = 2). Double the voltage (or halve the resistance), and the current doubles. Figure 10-4 shows this linear (straight-line graph) relationship.

Another form of Ohm's law is

$$R = \frac{E}{I}$$

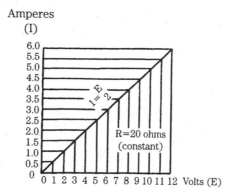

10-4 Relation between amperage and voltage, resistance constant.

or resistance equals voltage divided by current. The 12V potential of our hypothetical circuit delivers 2A, which means the resistance is 6Ω. The relationship between amperage and resistance with a constant voltage is illustrated in Figure 10-5.

Another way of expressing Ohm's law is

$$E = IR$$

or voltage equals current multiplied by resistance. Two amperes through 6Ω requires a potential of 12V. Double the resistance, and twice the voltage is required to deliver the same amount of current (Figure 10-6).

Various memory aids have been devised to help students remember Ohm's law. One involves an Indian, an eagle, and a rabbit. The Indian (I) sees the eagle (E) flying over the rabbit (R); this gives the relationship $I = E/R$. The eagle sees both the Indian and the rabbit in the same level, or $E = IR$. And the rabbit sees the eagle over the Indian, or $R = E/I$. A visual aid is shown in Figure 10-7. The circle is divided into three segments. To determine which form of the equation to use, put your finger on the quantity you want to solve for.

All of this might seem academic, and in truth, few mechanics perform calculations with Ohm's law. It is usually easier to measure all values directly with a meter. Still, it is very important to have an understanding of Ohm's law. It is the best description of simple circuits we have.

10-5 Relation between amperage and resistance, voltage constant.

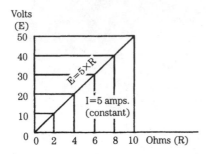

10-6 Relation between voltage and resistance, amperage constant.

For example, suppose the wiring is *shorted*—electrons have found a more direct (shorter) path to the positive terminal of the battery. Ohm's law tells us certain facts about the nature of shorts, which are useful in troubleshooting. First, a short increases the current in the affected circuit, because current values respond inversely to resistance. This increase will generate heat in the conductor and might even carbonize the insulation. Voltage readings will be low because the short has almost zero resistance. Now suppose we have a partially open circuit caused by a corroded terminal. The current through the terminal will be reduced, which means that the lights or whatever other load is on the circuit will operate at less-than-peak output. The terminal will be warm to the touch, because current is transformed to heat by the presence of resistance. Voltage readings from the terminal to ground will be high on the source side of the resistance and lower than normal past the bottleneck.

Direct and alternating current

Diesel engines might employ direct or alternating currents. The action of *direct,* or *unidirectional,* current can be visualized with the aid of the upper drawing in Figure 10-8. *Alternating current* is expressed by the opposed arrows in the lower drawing. Flow is from the negative to the positive poles of the voltage source, but the poles exchange identities, causing the current reversal. The positive pole becomes negative, and the negative pole becomes positive.

The graph in Figure 10-9 represents the rise, fall, and reversal of alternating current. Because the amplitude changes over time, alternating voltage and current values require some qualification. The next drawing illustrates the three values most often used.

Peak-to-peak values refer to the maximum amplitude of the voltage and amperage outputs in both directions. In Figure 10-10 the peak-to-peak value is 200 $V_{p\text{-}p}$ (or 200 $A_{p\text{-}p}$). The half-cycle (alternation) on top of the zero reference line is considered to be positive; below the line alternation is negative.

10-7 Ohm's law: Cover the unknown with your finger to determine which of the three equations to use.

Direct current

Alternating current

10-8 Direct and alternating current.

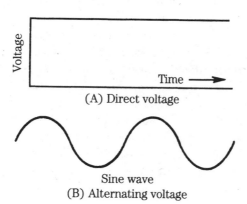

(A) Direct voltage

Sine wave
(B) Alternating voltage

10-9 Direct- and alternating-current waveforms.

70.7% — Effective

63.7% — Average

Peak to peak

Peak

10-10 Various values used to indicate sine-wave amplitude.

The *average,* or *mean,* value represents an average of all readings. In Figure 10-10 it is 63.7 units.

The *root mean square* (rms) value is sometimes known as the *effective* value. One rms ampere has the same potential for work as 1A direct current. Unless otherwise specified, alternating current values are *rms* (effective) values and are directly comparable to direct current in terms of the work they can do. Standard meters are equipped with appropriate scales to give rms readings.

The illustration depicts one complete cycle of alternating current. The number of cycles completed per second is the frequency of the current. One cycle per second is the same as one *hertz* (1 Hz). Domestic household current is generated at 60 Hz. The alternating current generators used with diesel engines are variable-frequency devices because they are driven by the engine crankshaft. At high speed a typical diesel alternating current generator will produce 500–600 Hz.

In addition to these special characteristics, alternating current outputs can be super-imposed upon each other (see Figure 10-11). The horizontal axis (X-axis) represents zero output—it is the crossover point where alternations reverse direction. In this par-ticular alternator 360° of rotation of the armature represents a complete out-put cycle from maximum positive to maximum negative and back to maximum positive. The X-axis can be calibrated in degrees or units of time (assuming that the alternator oper-ates at a fixed rpm). Note that there are two sine waves shown. These waves are out of phase; E_1 leads E_2 by 90° of alternator rotation. Alternators used on diesel installations usually generate three output waves 120° apart. Multiphase outputs are smoother than single-phase outputs and give engineers the opportunity to build multiple charging

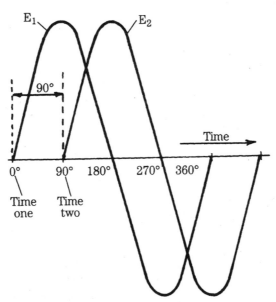

10-11 Voltage sine waves 90° apart.

circuits into the alternator. Circuit redundancy gives assurance against total failure; should one charging loop fail, the two others continue to function.

Magnetism

Electricity and magnetism are distinct phenomena, but related in the sense that one can be converted into the other. Magnetism can be employed to generate electricity, and electricity can be used to produce motion through magnetic attraction and repulsion.

The earth is a giant magnet with magnetic poles located near the geographic poles (Figure 10-12). Magnetic *fields* extend between the poles over the surface of the earth and through its core. The field consists of *lines of force*, or *flux*. These lines of force have certain characteristics, which, although they do not fully explain the phenomenon, at least allow us to predict its behavior. The lines are said to move from the north to the south magnetic pole just as electrons move from a negative to a positive electrical pole. Magnetic lines of force make closed loops, circling around and through the magnet. The poles are the interface between the internal and external paths of the lines of force. When encountering a foreign body, the lines of force tend to stretch and snap back upon themselves like rubber bands. This characteristic is important in the operation of generators.

Lines of force penetrate every known substance, as well as the emptiness of outer space. However, they can be deflected by soft iron. This material attracts and focuses flux in a manner analogous to the action of a lens on light. Lines of force "prefer" to travel through iron; they digress to take advantage of the *permeability* (magnetic conductivity) of iron (Figure 10-13).

The permeability of iron is exploited in almost all magnetic machines. Generators, motors, coils, and electromagnets all have iron cores to direct the lines of force most efficiently.

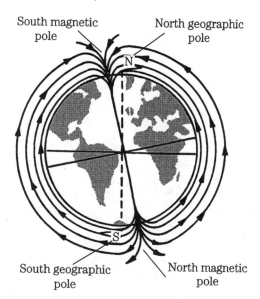

South magnetic pole

North geographic pole

South geographic pole

North magnetic pole

10-12 Earth's magnetic field.

10-13 Effects of soft iron on a magnetic field.

When free to pivot, a magnet aligns itself with a magnetic field, as shown in Figure 10-14. *Unlike magnetic poles attract, like magnetic poles repel.* The south magnetic pole of the small magnet (compass needle) points to the north magnetic pole of the large bar magnet.

Electromagnets

Electromagnets use electricity to produce a magnetic field. Electromagnets are used in starter solenoids, relays, and voltage-and-current-limiting devices. The principle upon which these magnets operate was first enunciated by the Danish researcher Hans Christian Oersted. In 1820, he reported a rather puzzling phenomenon—a compass needle when placed near a conductor, deflected as the circuit was made and broken. Subsequently it was shown that the needle reacted because of the presence of a magnetic field at right angles to the conductor. The field exists as long as current flows, and its intensity is directly proportional to the amperage.

The field around a single strand of conductor is too weak to have much by way of practical application. But if the conductor is wound into a coil, the weak fields of the turns reinforce each other (Figure 10-15). Further reinforcement can be obtained

Bar magnet

10-14 Compass deflections in a magnetic field.

10-15 Electromagnet construction showing how magnetic fields mutually reinforce one another.

by inserting an iron bar inside the coil to give focus to the lines of force. The strength of the electromagnet depends on the number of turns of the conductor, the length/width ratio of the coil, the current strength, and the permeability of the core material.

Voltage sources

Starting and charging systems employ two voltage sources. The generator is electromagnetic in nature and is the primary source. The battery, which is electrochemical in nature, is carried primarily for starting. Also, in case of overloads, the battery can send energy into the circuit.

Generator principles

In 1831 Michael Faraday reported that he had induced electricity in a conductor by means of magnetic action. His apparatus consisted of two coils, wound over each other but not electrically connected. When he made and broke the circuit to one coil, momentary bursts of current were induced in the second. Then Faraday wound the coils over a metal bar and observed a large jump in induced voltage. Finally he reproduced the experiment with a permanent magnet. Moving a conductor across a magnetic field produced voltage. The intensity of the voltage, depended on the strength of the field, the rapidity of movement, and the angle of movement. His generator was most efficient when the conductor cut the lines of force at right angles.

The rubber band effect mentioned earlier helps one to visualize what happens when current is induced in the conductor. A secondary magnetic field is set up, which resists the movement of the conductor through the primary field. The greater the amount of induced current, the greater the resistance, and the more power required to overcome it.

10-16 Voltage (EMF or electromotive force) produced by magnetism.

Alternator

Figure 10-16 illustrates Faraday's apparatus. Note that there must be relative movement between the magnetic field and conductor. Either can be fixed as long as one can move. Note also that the direction of current is determined by the direction of movement. Of course, this shuttle generator is hardly efficient; the magnet or conductor must be accelerated, stopped, and reversed during each cycle.

The next step was to convert Faraday's laboratory model to rotary motion (Figure 10-17). The output alternates with the position of the armature relative to the fixed magnets. In Figure 10-17A the windings are parallel to the lines of force and the output is zero. Ninety degrees later the output reaches its highest value because the armature windings are at right angles to the field. At 180° of rotation (C), the output is again zero. This position coincides with a polarity shift. For the remainder of the cycle, the movement of the windings relative to the field is reversed.

To determine the direction of current flow from a generator armature, position the left hand so that the thumb points in the direction of current movement and the forefinger points in the direction of magnetic flux (from the north to the south pole). Extend the middle finger 90° from the forefinger; it will point to the direction of current. As convoluted as the lefthand rule for generators sounds, it testifies to the fact that polarity shifts every 180° of armature rotation. This is true of all dynamos, whether the machine is DC generator or AC generator (alternator).

The frequency in hertz depends on the rotational speed of the armature and on the number of magnetic poles. Figure 10-18 illustrates a four-pole (two-magnet) alternator, which for a given rpm, has twice the frequency of the single-pole machine in

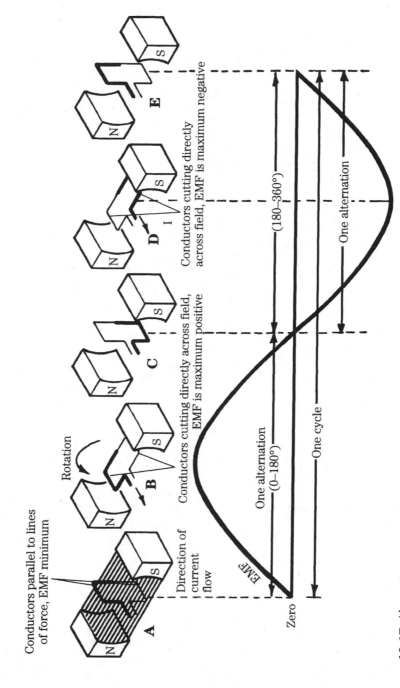

Conductors parallel to lines of force, EMF minimum

Rotation

Direction of current flow

Conductors cutting directly across field, EMF is maximum positive

Conductors cutting directly across field, EMF is maximum negative

One alternation (0–180°)

One cycle

Zero

EMF

(180–360°)

One alternation

A

B

C

D

E

I

10-17 Alternating current generator (or alternator).

10-18 Four-pole alternator.

the previous illustration. Normally, alternators supplied for diesel engines have four magnetic poles and are designed to peak at 6000 rpm. Their frequency is given by:

$$f = \frac{P \times \mathrm{rmp}}{120}$$

where f is frequency in hertz and P is the number of poles. Thus, at 6000 rpm, a typical alternator should deliver current at a frequency of 2000 Hz.

The alternators depicted in Figures 10-17 and 10-18 are simplifications. Actual alternators differ from these drawings in two important ways. First, the fields are not fixed, but rotate as part of an assembly called the *rotor*. Secondly, the fields are not permanent magnets; instead they are electromagnets whose strength is controlled by the voltage regulator. Electrical connection to the rotor is made by a pair of brushes and slip rings.

DC generator

Direct-current (DC) generators develop an alternating current (AC) that is mechanically rectified by the commutator—brush assembly. The commutator is a split copper ring with the segments insulated from each other. The brushes are carbon bars that are spring-loaded to bear against the commutator. The output is pulsating direct current, as shown in Figure 10-19. The pulse frequency depends on the number of armature loops,

10-19 Single-loop direct current generator.

field poles, and rpm. A typical generator has 24–28 armature loops and 4 poles. The fields are electromagnets and are energized by 8–12% of the generator output current. The exact amount of current detailed for this *excitation* depends on the regulator, which in turn, responds to the load and state of charge of the generator. A few surviving generators achieve output regulation by means of a third brush.

Third-brush generator

The third brush is connected to the field coils as shown in Figure 10-20. At low rotational speeds the magnetic lines of force bisect the armature in a uniform manner. But as speed and output increase, the field distorts. The armature generates its own field, because a magnetic field is created at right angles to a conductor when current flows through it. The resulting field distortion, sometimes called *magnetic whirl*, places the loops feeding the third brush in an area of relative magnetic weakness. Consequently, less current is generated in these loops, and the output to the fields is lessened. The fields become correspondingly weak, and the total generator output as defined by the negative and positive brushes remains stable or declines.

The output depends on the position of the third brush, which can be moved in or out of the distorted field to adjust output for anticipated loads. Moving the brush in the direction of armature rotation increases the output; moving it against the direction of rotation reduces the output. When operated independently from the external circuit, the brushes must be grounded to protect the windings. Many of these generators have a fuse in series with the fields.

Direct-current motors

One of the peculiarities of a direct-current generator is that it will "motor" if the brushes are connected to a voltage source. In like manner a direct current motor will "gen" if the armature is rotated by some mechanical means. Some manufacturers of

10-20 Third-brush generator. As speed and output increase, the field's excitation current drops because of magnetic whirl.

small-engine accessories have taken advantage of this phenomenon to combine both functions in a single housing. One such system is the *Dynastart* system employed on many single-cylinder Hatz engines (shown schematically in Figure 10-21).

Figure 10-22 illustrates the motor effect. In the top sketch the conductor is assumed to carry an electric current toward you. The magnetic flux around it travels in a clockwise direction. Magnetic lines of force above the conductor are distorted and stretched. Because the lines of force have a strong elastic tendency to shorten, they push against the conductor. Placing a loop of wire in the field (bottom sketch) instead

10-21 Combination generator and starter motor. Teledyne Wisconsin Motor

of a single conductor doubles the motor effect. Current goes in the right half of the loop and leaves at the left. The interaction of the fields causes the loop (or, collectively, the armature) to turn counterclockwise. Reversing the direction of the current would cause torque to be developed in a clockwise direction.

Starter motors are normally wired with the field coils in series with the armature (Figure 10-23). Any additional load added to a series motor will cause more current in the armature and correspondingly more torque. Because this increased current must pass through the series field, there will be a greater flux. Speed changes rapidly with load. When the rotational speed is low, the motor produces its maximum torque. A starter might draw 300A during cranking, more than twice that figure at stall.

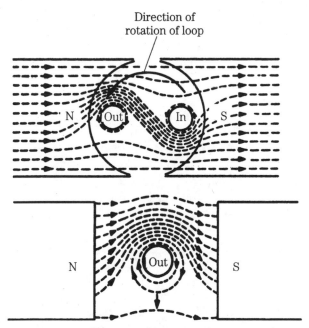

10-22 Motor effect created by current bearing conductor in a magnetic field.

10-23 Series field and armature connections typical of starter motors.

Storage batteries

Storage batteries accumulate electrical energy and release it on demand. The familiar lead-acid battery was invented more than 100 years ago by Gaston Plante. It suffers from poor energy density (watt-hours per pound) and poor power density (watts per pound). The average life is said to be in the neighborhood of 360 complete charge-discharge cycles. During charging, the lead-acid battery shows an efficiency of about 75%; that is, only three-quarters of the input can be retrieved.

Yet it remains the only practical alternative for automotive, marine, and most stationary engine applications. Sodium-sulfur, zinc-air, lithium-halide, and lithium-chlorine batteries all have superior performance, but are impractical by reason of cost and, in some cases, the need for complex support systems.

The lead-acid battery consists of a number of cells (hence the name *battery*) connected in series. Each fully charged cell is capable of producing 2.2V. The number of cells fixes the output; 12V batteries have six cells; 24V batteries have 12. The cells are enclosed in individual compartments in a rubberoid or (currently) high-impact plastic case. The compartments are sealed from each other and, with the exception of Delco and other "zero maintenance" types, open to the atmosphere. The lower walls of the individual compartments extend below the plates to form a sediment trap. Filler plugs are located on the cover and can be combined with wells or other visual indicators to monitor the electrolyte level.

The cells consist of a series of lead plates (Figure 10-24) connected by internal straps. In the past, these straps were routed over the top of the case, making convenient test points for the technician. Unfortunately, these straps leaked current and were responsible for the high self-discharge rates of these batteries.

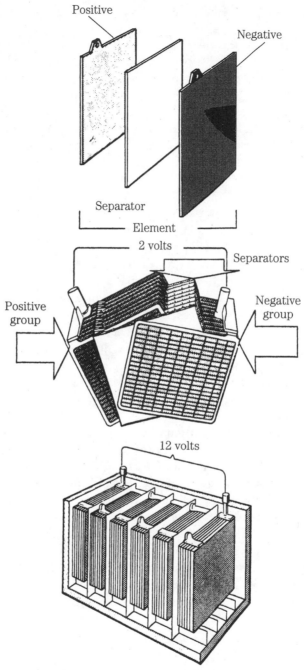

Positive

Negative

Separator

Element

2 volts

Separators

Positive group

Negative group

12 volts

10-24 Lead-acid storage battery construction.

The plates are divided into positive and negative groups and separated by means of plastic or fiberglass sheeting. Some very large batteries, which are built almost entirely by hand, continue to use fir or Port Orford cedar separators. A few batteries intended for vehicular service feature a loosely woven fiberglass padding between the separators and positive plates. The padding gives support to the lead filling and reduces damage caused by vibration and shock.

Both sets of plates are made of lead. The positive plates consist of a lead grid-work that has been filled with lead oxide paste. The grid is stiffened with a trace of antimony. Negative plates are cast in sponge lead. The plates and separators are immersed in a solution of sulfuric acid and distilled water. The standard proportion is 32% acid by weight.

The level of the electrolyte drops in use because of evaporation and hydrogen loss. (Sealed batteries have vapor condensation traps molded into the roof of the cells.) It must be periodically replenished with distilled water. Nearly all storage batteries are shelved dry, and filled upon sale. Once the plates are wetted, electrical energy is stored in the form of chemical bonds.

When a cell is fully charged the negative plates consist of pure sponge lead (Pb in chemical notation), and the positive plates are lead dioxide (PbO_2, sometimes called lead peroxide). The electrolyte consists of water (H_2O) and sulfuric acid (H_2SO_4). The fully charged condition corresponds to drawing A in Figure 10-25. During discharge both sponge lead and lead dioxide become lead sulfate ($PbSO_4$). The percentage of water in the electrolyte increases because the SO_4 radical splits off from the sulfuric acid to combine the plates. If it were possible to discharge a lead-acid battery completely, the electrolyte would be safe to drink. In practice, batteries cannot be completely discharged in the field. Even those that have sat in junkyards for years still have some charge.

During the charge cycle the reaction reverses. Lead sulfate is transformed back into lead and acid. However, some small quantity of lead sulfate remains in its crystalline form and resists breakdown. After many charge-discharge cycles, the residual sulfate permanently reduces the battery's output capability. The battery then is said to be *sulfated*.

The rate of self-discharge is variable and depends on ambient air temperature the cleanliness of the outside surfaces of the case, and humidity. In general it averages 1% a day. Batteries that have discharged below 70% of full charge might sulfate. Sulfation becomes a certainty below the 70% mark. In addition to the possibility of damage to the plates, a low charge brings the freezing point of the electrolyte to near 32°F.

Temperature changes have a dramatic effect on the power density. Batteries are warm-blooded creatures and perform best at room temperatures. At 10°F the battery has only half of its rated power.

Switches

Switches neither add energy to the circuit nor take it out. Their function is to give flexibility by making or breaking circuits or by providing alternate current paths. They are classified by the number of *poles* (movable contacts) and *throws* (closed positions). Figure 10-26 illustrates four standard switch types in schematic form.

10-25 Chemical action in a lead-acid cell.

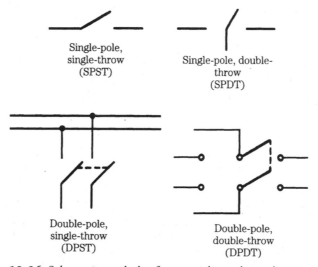

10-26 Schematic symbols of commonly used switches.

A single-pole, single-throw switch is like an ordinary light switch in that it makes and breaks a single circuit with one movement from the rest position. Single-pole, double-throw (SPDT) switches have three terminals, but control a single circuit with each throw. As one is completed, the other is opened. You can visualize an SPDT as a pair of SPST switches in tandem. Double-pole, double-throw switches control two circuits with a single movement. Think of the DPDT as two DPST switches combined, but working so that when one closes the other opens.

Self-actuating switches

Not all switching functions are done by hand. Some are too critical to trust to the operator's alertness. In general two activating methods are used. The first depends on the effect of heat on a bimetallic strip or disc to open or close contacts. Heat can be generated through electrical resistance or, as is more common, from, the engine coolant. The bimetallic element consists of two dissimilar metals bonded back-to-back. Because the coefficient of expansion is different for the metals, the strip or disc will deform inward, in the direction of the least expansive metal. Figure 10-27 illustrates this in a flasher unit. Other switches operate by the pressure of a fluid acting against a diaphragm. The fluid can be air, lube oil, or brake fluid. The most common use of such switches is as oil pressure sensors (Figure 10-28).

Relays

Relays are electrically operated switches. They consist of an electromagnet, a movable armature, a return spring, and one or more contact sets. Small currents excite the coil, and the resulting magnetic field draws the armature against the coil.

10-27 Bimetallic switch operation.

Lamp ON when no oil pressure is available   Lamp OFF when oil pressure is available

10-28 Diaphragm switch (oil pressure).

This linear movement closes or opens the contacts. The attractiveness of relays is that small currents—just enough to excite the electromagnet—can be used to switch large currents. You will find relays used to activate charging indicator lamps (Figure 10-29), some starter motors, horns, remotely controlled air trips, and the like.

Solenoids are functionally similar to relays. The distinction is that the movable armature has some nonelectrical chore, although it might operate switch contacts in the bargain. Many starter motors employ a solenoid to lever the pinion gear into engagement with the flywheel. The next chapter discusses several of these starters.

Circuit protective devices

Fuses and fusible links are designed to vaporize and open the circuit during overloads. Some circuits are routed through central fuse panels. Others can be protected by individual fuses spliced into the circuit at strategic points.

Fusible links are used as a protection for the battery-regulator circuit (Figure 10-30). Burnout is signaled by swollen or discolored insulation over the link. New links can be purchased in bulk for soldered joints, or in precut lengths for use with quick-disconnects. Before replacing the link, locate the cause of failure, which will be a massive short in the protected circuit.

10-29 Relay-tripped charging indicator lamp.

10-30 Fusible link.

On the other hand, a blown fuse is no cause for alarm. Fuses fatigue and become resistive with age. And even the best-regulated charting circuit is subject to voltage and current spikes, which can take out a fuse but otherwise are harmless. Chronic failure means a short or a faulty regulator.

Capacitors (condensers)

Two centuries ago it was thought that Leyden jars for storing charges actually condensed electrical "fluid." These glass jars, wrapped with foil on both the inner and outer surfaces, became known as *condensers*. The term lingers in the vocabulary of automotive and marine technicians, although *capacitor* is more descriptive of the devices' capacity for storing electricity.

A capacitor consists of two plates separated by an insulator or *dielectric* (Figure 10-31). In most applications the plates and dielectric are wound on each other to save space.

A capacitor is a kind of storage tank for electrons. In function it is similar to an accumulator in hydraulic circuits. Electrons are attracted to the negative plate by the close proximity of the positive plate. Storing the electrons requires energy, which is

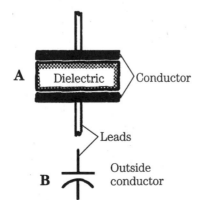

10-31 Capacitor construction and schematic symbol.

released when the capacitor is allowed to discharge. Shorting the plates together sends free electrons out of the negative plate and to the positive side. There is a to-and-fro shuttle of electrons between the plates until the number of electrons on the plates is equalized. The number of oscillations depends on circuit resistance and *reactance* (or the reluctance of the capacitor to charge because of the mutual repulsion of electrons on the negative plate).

Capacitors block direct current, but pass alternating current. The electrons that make up alternating current do not physically penetrate the dielectric, but their shifting polarity attracts or repulses electrons clustered on the other side of the dielectric. High-frequency AC behaves as if the dielectric did not exist.

Capacitors slowly degrade in service or, should the dielectric puncture, fail catastrophically. Those capacitors that are not buried in integral circuits and accessible for replacement are pretty well limited to the output side of alternators, where they divert high-frequency radio interference to ground. Substitution is the most reliable way of testing capacitors, although an ohmmeter will detect gross, zero-resistance dielectric failure.

Diodes

Diodes are solid-state components that act like traffic cops to block current moving in one direction and pass current in the other direction. They are used in voltage regulators and in charging circuits to rectify AC alternator output to pulsating DC suitable for battery charging. When tested with an ohmmeter, a diode should have low resistance with the test leads connected in one way, and high, almost infinite resistance with the leads are reversed. Replacement diodes cost less if purchased from an electronics supply house. Just make sure the new part has the same or better current and voltage ratings as the original. Diodes pressed into aluminum heat sinks may require fabricating a tool for installation. A coat of dielectric grease assists in heat transfer.

Diodes are artificial silicon or germanium crystals, grown in the laboratory. During the growth stage, the crystals are "doped" with precise levels of indium, aluminum or boron. These impurities free electrons from the crystalline structure to produce what is called *n*-material (*n* for negative). Doping with arsenic or phosphorus creates a shortage of electrons in what is called *p*, or positive material. The missing electrons, or "holes," await replenishment by electrons with a negative charge. These two doped crystals—*p*-material and *n*-material—are joined in a junction.

Applying a negative voltage to the junction forces electrons out of the negative material that cross the junction and fill the holes in the positive material. Like charges repel and unlike charges attract. A positive voltage, one the other hand, pulls electrons from the *n*-material and the *p*-material migrates away from the junction. No current passes through the diode.

Why diodes fail

The failure mode is absolute. Either the diode works or it doesn't. Failure might be spontaneous—the result of manufacturing error or it might be the result of faulty service procedures. Service-caused failures increase over time as opportunities for human error multiply.

Service-related failures are caused by:

High inverse voltage resulting from wrong polarity Jumper cables connected backward or a battery installed wrong will scramble the crystalline structure.

Vibration and mechanical shock Diodes must be installed in their heat sinks with the proper tools. A 2 × 4 block is not adequate.

Short circuits The high levels of current flow that result from short circuits overheat diodes.

Soldering The standard practice is for alternator diodes to be soldered to the stator leads. Too much heat at the connection can destroy the diodes.

11
CHAPTER

Starting and generating systems

The diesel engines we are concerned with are almost invariably fitted with electric starter motors. A number of engines, used to power construction machinery and other vehicles that are expected to stand idle in the weather, employ a gasoline engine rather than an electric motor as a starter. Stationary engines are sometimes fitted with compressed-air starter motors.

Starting a cold engine can be somewhat frustrating, particularly if the engine is small. The surface/volume ratio of the combustion space increases disproportionately as engine capacity is reduced. The heat generated by compression tends to dissipate through the cylinder and head metal. In addition, cold clearances might be such that much of the compressed air escapes past the piston rings. Other difficulties include the effect of cold on lube and fuel oil viscosity. The spray pattern coarsens, and the drag of heavy oil between the moving parts increases.

Starting has more or less distinct phases. Initial or breakaway torque requirements are high because the rotating parts have settled to the bottom of their journals and are only marginally lubricated. The next phase occurs during the first few revolutions of the crankshaft. Depending on ambient temperature, piston clearances, lube oil stability, and the like, the first few revolutions of the crankshaft are free of heavy compressive loads. But cold oil is being pumped to the journals, which collects and wedges between the bearings and the shafts. As the shafts continue to rotate, the oil is heated by friction and thins, progressively reducing drag. At the same time, cranking speed increases and compressive loads become significant. The engine accelerates to *firing speed*. The duration from breakaway to firing speed depends on the capacity of the starter and battery, the mechanical condition of the engine, lube oil viscosity, ambient air temperature, the inertia of the flywheel, and the number of cylinders. A single-cylinder engine is at a disadvantage because it cannot benefit from the expansion of other cylinders. Torque demands are characterized by sharp peaks.

Starting aids

It is customary to include a cold-starting position at the rack. This position provides extra fuel to the nozzles and makes combustion correspondingly more likely.

Lube oil and water immersion heaters are available that can be mounted permanently on the engine. Lube oil heaters are preferred and can be purchased from most engine builders. Good results can be had by heating the oil from an external heater mounted below the sump. Use an approved type to minimize the fire hazard. Alternatively, one can drain the oil upon shutdown and heat it before starting. The same can be done with the coolant, although temperatures in both cases should be kept well below the boiling temperature of water to prevent distortion and possible thermal cracking.

If extensive cold weather operation is intended or if the engine will be stopped and started frequently, it is wise to add one or more additional batteries wired in parallel. Negative-to-negative and positive-to-positive connections do not alter the output voltage, but add the individual battery capacities.

Once chilled beyond the cloud point, diesel fuel enters the gelling stage. Flow through the system is restricted, filter efficiency suffers, and starting becomes problematic. Racor is probably the best known manufacturer of fuel heaters, which are available in a variety of styles. Several combine electric resistance elements with a filter, to heat the fuel at the point of maximum restriction. Another type incorporates a resistance wire in a flexible fuel line.

Makers of indirect injection engines generally fit glow plugs as a starting aid (Figure 11-1). These engines would be extremely difficult to start without some method of heating the air in the prechamber. A low-resistance filament (0.25–1.5 Ω, cold) draws heavy current to generate 1500°F at the plug tip. Early types used exposed filaments, which sometimes broke off and became trapped between the piston and chamber roof with catastrophic effects on the piston and (when made of aluminum) the head. Later variants contain the filament inside of a ceramic cover, which eliminates the problem. However, ceramic glow plugs are quite vulnerable to damage when removed from the engine and must be handled with extreme care.

In all cases, glow plugs are wired in parallel and controlled by a large power relay. Test filament continuity with an ohmmeter.

Primitive glow-plug systems are energized by a switch, sometimes associated with a timer, and nearly always in conjunction with a telltale light. The more sophisticated systems used in contemporary automobiles automatically initiate glow-plug operation during cranking and, once the engine starts, gradually phase out power.

Heating body   Terminal

11-1 Sheathed-type glow plug. Marine Engine Div., Chrysler Corp.

Two types of circuits are encountered, both built around a solid-state module with an internal clock. The pulsed system opens the glow-plug power circuit for progressively longer intervals as the engine heats and the timer counts down. In the Ford/Navistar version of this circuit, glow-plug resistance varies with tip temperature, so that the plugs themselves function as heat sensors. Note that these low-resistance devices self-destruct within seconds of exposure to steady-state battery voltage. Pulsed glow plugs can be tested with a low-voltage ohmmeter and plug operation can be observed by connecting a test lamp between the power lead and the glow-plug terminal. Normally, if the circuit pulses, it can be considered okay; when in doubt, consult factory literature for the particular engine model.

Most manufacturers take a less ambitious approach, and limit glow-plug voltage by switching a resistor into the feed circuit. During cold starts, a relay closes to direct full battery voltage to the glow plugs; as the engine heats (a condition usually sensed at the cylinder-head water jacket), the first relay opens and a second relay closes to switch in a large power resistor. Power is switched off when the module times out.

Starting fluid can be used in the absence of intake air heaters. In the old days a mechanic poured a spoonful of ether on a burlap rag and placed it over the air intake. This method is not the safest nor the most consistent; too little fluid will not start the engine, and too much can cause severe detonation or an intake header explosion. Aerosol cans are available for injection directly into the air intake. Use as directed in a well-ventilated place.

More sophisticated methods include pumps and metering valves in conjunction with pressurized containers of starting fluid. Figure 11-2 illustrates a typical metering valve. The valve is tripped only once during each starting attempt, to forestall explosion. Caterpillar engines are sometimes fitted with a one-shot starting device consisting of a holder and needle. A capsule of fluid is inserted in the device and the needle pierces it, releasing the fluid.

The starter motor should not be operated for more than a few seconds at a time. Manufacturers have different recommendations on the duration of cranking, but none suggests that the starter button be depressed for more than 30 seconds. Allow a minute or more between bouts for cooling and battery recovery.

Wiring

Figure 11-3 illustrates a typical charging/starting system/in quasi-realistic style. The next drawing (Figure 11-4) is a true schematic of the same system, encoded in a way that conveys the maximum amount of information per square inch.

Neither of these drawings is to scale and the routing of wires has been simplified. The technician needs to know where wires terminate, not what particular routes they take to get there. When routing does become a factor, as in the case of electronic engine control circuitry, the manufacturer should provide the necessary drawings.

Fluid
cylinder

Actuator
cable

Metering
valve

Valve
lever

11-2 Quick-Start Unit. Detroit Diesel

Color coding

As is customary, the schematic in Figure 11-4 indicates the color of the insulation
and the size of the various conductors. Color coding has not been completely stan-
dardized, but most manufacturers agree that black should represent ground. This does
not mean that colors change in a purely arbitrary fashion. In the example schematic,
red denotes the positive side of the battery (the "hot" wire); white/blue is associated
with the temperature switch; orange with the tachometer, and so on.

Two-color wires consist of a base, or primary, color and a tracer. The base color
is always first in the nomenclature. Thus, white/blue means a wire with a blue stripe.

Cooling water
temperature switch

AC alternator

Starter - motor

Fuse box
(30A)

Battery

Tachometer
sensor

Lubricating oil
pressure switch

Battery switch
(optional)

Extension
wireharness

For GM series:
12V, 70A-hr or higher
For Model 3HM Series.
12V, 100A-hr or higher

Wireharness
instrument panel

B-type instrument panel (large)

A-type instrument panel (small)

Rubber seated lamp (sail-drive)

Water temperature warning lamp

Oil pressure warning lamp

Charge lamp

Tachometer

Alarm buzzer

Lamps switch
for tachometer

Push button switch

Key switch

Rubber sealed lamp (sail-drive)

Water temperature
warning lamp

Oil pressure
warning lamp

Charge
lamp

Push button
switch

Alarm buzzer

Key switch

11-3 Wiring layout for a Yanmar Marine application.

Wiring repairs

The material that follows applies to simple DC starting, charging, and instrumentation circuits. Cutting or splicing wiring harnesses used with electronic engine management systems can do strange things to the computer.

The first consideration when selecting replacement wire is its current-carrying capacity, or gauge. In the context of diesel engines, two more or less interchangeable

11-4 A detailed schematic for the wiring layout shown previously.

Table 11-1. SAE/metric wire sizes

SAE wire size	Sectional area		Outer dia.		Conductor resistance (Ω/km)	Allowable continuous load current (A)
	Circular mils	mm²	in.	mm		
20	1,094	0.5629	0.040	1.0	32.5	7
18	1,568	0.8846	0.050	1.2	20.5	9
16	2,340	1.287	0.060	1.5	14.1	12
14	3,777	2.091	0.075	1.9	8.67	16
12	5,947	3.297	0.090	2.4	5.50	22
10	9,443	5.228	0.115	3.0	3.47	29
8	15,105	7.952	0.160	3.7	2.28	39
6	24,353	13.36	0.210	4.8	1.36	88
4	38,430	20.61	0.275	6.0	0.871	115
2	63,119	35.19	0.335	8.0	0.510	160
1	80,010	42.73	0.375	8.6	0.420	180
0	100,965	54.29	0.420	9.8	0.331	210
2/0	126,882	63.84	0.475	10.4	0.281	230
3/0	163,170	84.96	0.535	12.0	0.211	280
4/0	207,740	109.1	0.595	13.6	0.135	340

Source: Marine Engine Div., Chrysler Corp.

standards apply—the Society of Automotive Engineers (SAE) and Japanese Industrial Standard (JIS).

For most wire, the smaller the SAE gauge number, the greater the cross-sectional area of the conductor (See Table 11.1). Thus, #10 wire will carry more current than #12. The schema reverses when we get into heavy cable: 4/0 is half again as large as 2.0 and has a correspondingly greater current capacity. Note that one must take these current values on faith.

The JIS standard eliminates much confusion by designating wire by type of construction and cross-sectional area. Thus, JIS AV5 translates as automotive-type (stranded) wire with a nominal conductor area of 5 mm². The Japanese derive current-carrying capacity from conductor temperature, which cannot exceed 60°C (140°F). Ambient temperature and wire type affect the rating, as shown back in Table 4-2.

Vinyl-insulated, stranded copper wire is standard for engine applications. Teflon insulation tolerates higher temperatures than vinyl and has better abrasion resistance. But Teflon costs more and releases toxic gases when burned. In no case should you use Teflon-insulated wire in closed spaces.

All connections should be made with terminal lugs. Solder-type lugs (Figure 11-5A) can provide mechanically strong, low-resistance joints and are infinitely preferable to the crimp-on terminals shown in Figure 11-5B. Insulate with shrink tubing. When shrink tubing is impractical (as when insulating a Y-joint), use a good grade of vinyl electrician's tape. The 3-M brand costs three times more than the imported variety and is worth every penny.

Table 11-2. Allowable amperage and voltage drop (JIS AV wire)

Ambient temp.	30°C		40°C		50°C	
Allowable amperage/ Voltage drop	A	mv/m	A	mv/m	A	mv/m
Nominal section (mm²)						
0.5	11	414	9	338	6	226
0.85	15	356	12	284	8	190
1.25	19	310	15	245	10	163
2	25	251	20	201	14	140
3	34	216	28	178	19	121
5	46	185	37	149	26	104
8	60	158	49	129	34	90
15	82	129	66	104	46	72
20	109	110	86	87	61	62
30	152	90	124	73	87	51
40	170	83	139	68	98	48
50	195	75	159	61	113	43
60	215	70	175	57	124	40
85	254	62	207	51	146	36
100	294	56	240	46	170	33

Figure 11-6 illustrates how stranded wire is butt spliced. Cut back the insulation ¾ in. or so, and splay the strands apart. Push the wires together, so that the strands interleave, and twist. Apply a small amount of solder to the top of the joint, heating from below.

Soldering

Most mechanics believe they know how to solder, but few have received any training in the art. Applied correctly, solder makes a molecular bond with the base metal. Scrape conductors bright to remove all traces of oxidation.

11-5 (A) Solder-type terminal lug. (B) Crimp-on terminal lug.

11-6 Splice with stranded conductor (A) strands splayed, (B) interleaved, (C) twisted.

Use a good grade of low-temperature, rosin-core solder, such as Kessler "blue," which consists of 60% tin and 40% lead. A 250-W gun should be adequate for all but the heaviest wiring.

The tip of a nonplated soldering iron should be dressed with a file down to virgin copper and tinned, or coated with molten solder. Silver solder is preferable for tinning because it melts at a higher temperature than lead-based solder and so protects the tip from corrosion. Retighten the tip periodically.

The following rules were developed from experience and from a series of experiments conducted by the military:

- Use a minimum amount of solder.
- Wrapping terminal lugs and splicing ends with multiple turns of wire does not add to the mechanical strength of the joint and increases the heat requirement. Wrap only to hold the joint while soldering.
- Heat the connection, not the solder. When the parts to be joined are hot enough, solder will flow into the joint.
- Do not move the parts until the solder has hardened. Movement while the solder is still plastic will produce a highly resistive "cold" joint.
- Use only enough heat to melt the solder. Excessive heat can damage nearby components and can crystallize the solder.
- Allow the joint to air cool. Dousing a joint with water to cool it weakens the bond.

Starter circuits

Before assuming that the motor is at fault, check the battery and cables. The temperature-corrected hydrometer reading should be at least 1.240, and no cell should vary from the average of the others by more than 0.05 point. See that the battery terminal connections are tight and free of corrosion.

Excessive or chronic starter failure might point to a problem that is outside the starter itself. It could be caused by an engine that is out of tune and that consequently requires long cranking intervals.

Starter circuit tests

There are several methods that you can use to check the starting-circuit resistance. One method is to open all the connections, scrape bright, and retighten. Another method requires a low-reading ohmmeter of the type sold by Sun Electric and other suppliers for the automotive trades. But most mechanics prefer to test by voltage drop.

11-7 Hot side voltage drop test.

Connect a voltmeter as shown in Figure 11-7. The meter shunts the positive, or hot battery post and the starter motor. With the meter set on a scale above battery voltage, crank. Full battery voltage means an open in the circuit.

If the starter functions at all, the reading will be only a fraction of this. Expand the scale accordingly. A perfect circuit will give a zero voltage drop because all current goes to the battery. In practice some small reading will be obtained. The exact figure depends on the current draw of the starter and varies between engine and starter motor types. As a general rule, subject to modification by experience, a 0.5V drop is normal. Much more than this means: (1) resistance in the cable, (2) resistance in the connections (you can localize this by repeating the test at each connection point), or (3) resistance in the solenoid.

Figure 11-8 shows the connections for the ground-side check. A poor ground, and consequent high voltage on the meter, can occur at the terminals, the cable, or between the starter motor and engine block. If the latter is the case, remove the motor and clean any grease or paint from the mounting flange.

Starter motors

Starter motors are series-wound; i.e., they are wound so that current enters the field coils and goes to the armature through the insulated brushes. Because a series-wound motor is characterized by high no-load rpm, some manufacturers employ limiting coils in shunt with the fields. The effect is to govern the free-running rpm and prolong starter life should the starter be energized without engaging the flywheel.

The exploded view in Figure 11-9 illustrates the major components of a typical starter motor. The frame (No. 1) has several functions. It locates the armature and fields, absorbs torque reaction, and forms part of the magnetic circuit.

11-8 Ground circuit test.

The field coils (No. 2) are mounted on the pole shoes (No. 3) and generate a magnetic field, which reacts with the field generated in the armature to produce torque. The pole pieces are secured to the frame by screws.

The armature (No. 4) consists of a steel form and a series of windings, which terminate at the commutator bars. The shaft is integral with it and splined to accept the starter clutch.

11-9 Starter in exploded view and as assembled. Lehman Manufacturing Co., Inc.

The end plates (5 and 6) locate the armature by means of bronze bushings. The commutator end plate doubles as a mounting fixture for the brushes, while the power takeoff side segregates the starter motor from the clutch.

The insulated (hot) brushes (No. 7) provide a current path from the field coils through the commutator and armature windings to the grounded brushes (No. 8).

Engagement of this particular starter is done by means of a yoke (No. 9), which is pivoted by the solenoid plunger (No. 10) in response to current flowing through the solenoid windings (No. 11). Movement of the plunger also trips a relay (No. 12) and energizes the motor. The pinion gear (No. 13) meshes with the ring gear on the rim of the flywheel. The pinion gear is integral with an overrunning clutch.

The starter drive housing supports the power takeoff end of the shaft and provides an accurately machined surface for mounting the starter motor to the ending block or bellhousing.

Brushes

Before any serious work can be done, the starter must be removed from the engine, degreased, and placed on a clean bench. Disconnect one or both cables at the battery to prevent sparking; disconnect the cable to the solenoid and the other leads that might be present (noting their position for assembly later); and remove the starter from the flywheel housing. Starters are mounted with a pair of cap screws or studs.

Remove the brush cover, observing the position of the screw or snap, because wrong assembly can short the main cable or solenoid wire (Figure 11-10). Hitachi starters do not have an inspection band as such. The end plate must be removed for access to the brushes and commutator.

Brushes are sacrificial items and should be replaced when worn to half their original length. The rate of wear should be calculated so that the wear limit will not be reached between inspection periods. Clean the brush holders and commutator with a preparation intended for use on electrical machinery. If old brushes are used, lightly

11-10 CAV CA-45C starter; band location in inspection.
GM Bedford Diesel

11-11 Testing brush holder insulation. Tecumseh Products Co.

file the flanks at the contact points with the holders to help prevent sticking. New brushes are contoured to match the commutator, but should be fitted by hand. Wrap a length of sandpaper around the commutator—do not use emery cloth—and turn in the normal direction of rotation. Remove the paper and blow out the dust.

Try to move the holders by hand. Most are riveted to the end plate and can become loose, upsetting the brush-commutator relationship. With an ohmmeter, test the insulation on the hot-side brush holders (Figure 11-11). There should be no continuity between the insulated brush holders and the end plate. Brush spring tension is an important and often overlooked factor in starter performance. To measure it, you will need an accurate gauge such as one supplied by Sun Electric. Specifications vary between makes and models, but the spring tension measured at the free (brush) end of the spring should be at least 1½ lb. Some specifications call for 4 lb.

The commutator bars should be examined for arcing, scores, and obvious eccentricity. Some discoloration is normal. If more serious faults are not apparent, buff the bars with a strip of 000 sandpaper.

Armature

Further disassembly requires that the armature shaft be withdrawn from the clutch mechanism. Some starters employ a snap ring at the power takeoff end of the shaft to define the outer limit of pinion movement. Others use a stop nut (Figure 11-12, No. 49). The majority of armatures can be withdrawn with the pinion and overrunning clutch in place. The disengagement point is at the yoke (Figure 11-9, No. 9) and sleeve on the clutch body. Remove the screws holding the solenoid housing to the frame and withdraw the yoke pivot pin. The pin might have a threaded fastener with an eccentric journal, as in the case of Ford designs. The eccentric allows for wear compensation. Or it might be a simple cylinder, secured by a flanged head on one side and a cotter pin or snap ring on the other. When the pin is removed there will be enough slack in the mechanism to disengage the yoke from the clutch sleeve, and the armature can be withdrawn. Observe the position of shims—usually located between the commutator and end plate—and, on CAV starters, the spring-loaded ball. This mechanism is shown in Figure 11-12 as 18 and 34. It allows a degree of end float so that the armature can recoil if the pinion and flywheel ring gear do not mesh on initial contact.

The armature should be placed in a jig and checked for trueness because a bent armature will cause erratic operation and might, in the course of long use, destroy the flywheel ring gear. Figure 11-13 illustrates an armature chucked between lathe

Pinion
stop nut 49
Thrust
washer 48
Shim washer 47
Pinion-assy 46
Dust scraper
ring 45
Lubricator wick 44
Selenoid
plunger 43
2nd contact 42
Trigger spring 41
Resistor 40
Trigger 39
Lock collar 38

1 Helix cover
2 Pinion spring
3 Spring
Lubricator
4 core plug
Drive end
5 shield
Selenoid
6 assy
7 1st contact
8 Locking balls
9 Lock collar spring
10 Trip collar
11 Circlip
12 Pole screw
13 Pole shoes
14 Yoke

Commutator
cover 37
Shims 36
Throughbolts 35
Recoil spring 34
Thrust washer 33
Endcap 32

Commutator
15 end shield
16 Shim washers
17 Circlip
18 Steel ball
19 Insulating brush
20 Terminal nut
Solenoid
21 terminal
22 Main terminal
23 Terminal nut
24 Insulating brush
25 Screw

Self-tapping
screws 31
Field terminal
screw 30
Cover plug 29
Resistor
flexible lead 28
Resistor rivet 27
Spacer 26

11-12 Typical CAV starter. GM Bedford Diesel.

centers. The allowable deflection at the center bearing is 0.002 in., or 0.004 in. on the gauge. With the proper fixtures and skill with an arbor press, an armature shaft can be straightened, although it will not be as strong as it was originally and will be prone to bend again. The wiser course is to purchase a new armature.

Make the same check on the commutator. Allowable out-of-roundness is 0.012– 0.016 in., or less, depending on the rate of wear and the intervals between inspection periods. Commutators that have lost their trueness or have become pitted should be

11-13 Checking armature shaft defection. Marine Engine Div, Chrysler Corp.

turned on a lathe. The cutting tool must be racked more for copper than for steel. Do not allow the copper to smear into the slots between the segments. Chamfer the end of the commutator slightly. Small imperfections can be removed by chucking the commutator in a drill press and turning against a single-cut file.

It is necessary that the insulation (called, somewhat anachronistically, *mica*) be buried below the segment edges; otherwise, the brushes will come into contact with the insulation as the copper segments wear. Undercutting should be limited to 0.015 in. or so. Tools are available for this purpose, but an acceptable job can be done with a hacksaw blade (flattened to fit the groove) and a triangular file for the final bevel cut (Figure 11-14).

Inspect the armature for evidence of overheating. Extended cranking periods, dragging bearings, chronically low battery charge, or an under-capacity starter will cause the solder to melt at the commutator-armature connections. Solder will be splattered over the inside of the frame. Repairs might be possible because these connections are accessible. Continued overheating will cause the insulation to flake and powder. Discoloration is normal, but the insulation should remain resilient.

Armature insulation separates the nonferrous parts (notably the commutator segments) from the ferrous parts (the laminated iron segments extending to the outer diameter of the armature and shaft). Make three tests with the aid of a 120V continuity lamp or a *megger* (meter for very high resistances). An ordinary ohmmeter is useless to measure the high resistances involved. In no case should the lamp light up or resistance be less than 1 MΩ.

11-14 Undercutting mica. Marine Engine Div., Chrysler Corp.

Electrical source

Acts as shorted circuit

Ground

No electrical flow

Increased electical flow

Plug into 120 volt source

11-15 Continuity   check between commutator bars and armature segments.

WARNING: Exercise extreme care when using 120V test equipment.

- Test between adjacent commutator segments.
- Test between individual commutator segments and the armature form (Figure 11-15).
- Test between armature or commutator segments and the shaft.

It is possible for the armature windings to short, thus robbing starter torque. Place the armature on a growler and rotate it slowly while holding a hacksaw blade over it as shown in Figure 11-16. The blade will be strongly attracted to the armature segments because of the magnetic field introduced in the windings by the growler.

Electrical source

Increased electical flow

Plug into 120 volt source

11-16 Checking   for   shorts with a growler.

But if the blade does a Mexican hat dance over a segment, you can be sure that the associated winding is shorted.

Field coils

After armatures, the next most likely source of trouble is the field coils. Hookups vary. The majority are connected in a simple series circuit, although you will encounter starters with *split fields*, each pair feeding off its own insulated brush. Field resistance values are not as a rule supplied in shop manuals, although a persistent mechanic can obtain this and other valuable test data by contacting the starter manufacturer. Be sure to include the starter model and serial number.

In the absence of a resistance test, which would detect intracoil shorts, the only tests possible are to check field continuity (an ordinary ohmmeter will do) and to check for shorts between the windings and frame. Connect a lamp or megger to the fields and touch the other probe to the frame, as illustrated in Figure 11-17. Individual fields can be isolated by snipping their leads.

The fields are supported by the pole shoes, which in turn, are secured to the frame by screws. The screws more often than not will be found to have rusted to the frame. A bit of persuasion will be needed, in the form of penetrating oil and elbow grease. Support the frame on a bench fixture and, with a heavy hammer, strike the screwdriver exactly as if you were driving a spike. If this does not work, remove the screw with a cape chisel. However, you might (depending on the starter make and your supply of junk parts) have difficulty in matching the screw thread and head fillet.

Coat the screw heads with Loctite before installation and torque securely. Be sure that the pole shoe and coil clears the armature and that the leads are tucked out of the way.

Bearings

The great majority of starters employ sintered bronze bushings. In time these bushings wear and must be replaced to ensure proper teeth mesh at the flywheel and prevent armature drag. In extreme cases the bushings can wear down to their bosses so

Plug into 120 volt source

Field terminal

Starter housing

11-17 Checking for grounded fields. Tecumseh Products Co.

that the shaft rides on the aluminum or steel end plates. The old bushings are pressed out and new ones pressed in. Tools are available to make this job easier, particularly at the blind boss on the commutator end plate.

Without these tools, the bushing can be removed by carefully ridging and collapsing it inward, or by means of hydraulic pressure. Obtain a rod that matches shaft diameter. Pack the bushing with grease and hammer the rod into the boss. Because the grease cannot easily escape between the rod and bushing, it will lift the bushing up and out. Because of the blind boss, new bushings are not reamed.

Final tests

The tests described thus far have been *static* tests. If a starter fails a static test, it will not perform properly, but passing does not guarantee the starter is faultless. The only sure way to test a starter—or, for that matter, any electrical machine—is to measure its performance under known conditions, and against the manufacturer's specifications or a known-good motor.

No-load performance is checked by mounting the starter in a vise and monitoring voltage, current, and rpm. Figure 11-18 shows the layout. The voltage is the reference for the test and is held to 12V. Current drain and rpm are of course variable, depending on the resistance of the windings, their configuration, and whether or not speed-limiting coils are provided. You can expect speeds of 4000–7000 rpm and draws of 60–100A. In this test we are looking for low rpm and excessive current consumption.

The locked-rotor and stall test requires a scale to accept the pinion gear (Figure 11-19). It must be made quickly, before the insulation melts. Mount the motor securely and lock the pinion. Typically the voltage will drop to half the normal value. Draw can amount to several hundred amperes.

Solenoids

Almost all diesel starters are engaged by means of a solenoid mounted on the frame. At the same time the solenoid plunger moves the pinion, it closes a pair of contacts to complete the circuit to the starter motor. In other words the component consists of a solenoid or *linear motor* and a relay (Figure 11-20). Some circuits feature

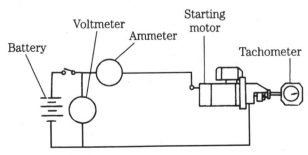

11-18 No-load starter test. Marine Engine Div., Chrysler Corp.

11-19 Stall torque test. Multiply scale reading by lever length in feet to obtain torque output. Onan

a second remotely mounted relay, as shown in Figures 11-20 and 11-21. The circuit depicted in the earlier illustration was designed for marine use.

The starter switch might be 30 ft or more from the engine. To cut wiring losses a second relay is installed, which energizes the piggyback solenoid. The additional relay (2nd relay in Figure 11-21) also gives overspeed protection should the switch remain closed after the engine fires. It functions in conjunction with a direct current generator. Voltage on the relay windings is the difference between generator output and battery terminal voltage. When the engine is cranking, generator output is functionally zero. The relay closes and completes the circuit to the solenoid. When the engine comes up to speed, generator output bucks battery voltage and the relay opens, automatically disengaging the starter. As admirable as such a device is, it should not be used continually, because some starter overspeeding will still occur, with detrimental effects on the motor bearings.

11-20 Solenoid internal wiring diagram.

11-21 Overspeed relay and solenoid wiring diagram. GM Bedford Diesel

Should the generator circuit open, the starter will be inoperative because the return path for the overspeed relay is through the generator. In an emergency the engine can be started by bridging the overspeed terminals. Of course, the transmission or other loads must be disengaged and, on a vehicle, the handbrake must be engaged.

In addition, you will notice that the solenoid depicted in Figure 11-21 has two sets of contacts. One set closes first and allows a trickle of current to flow through the resistor (represented by the wavy line above the contact arm) to the motor. The armature barely turns during the engagement phase. But once engaged a trigger is released and the second set of contacts closes, shunting the resistor and applying full battery current to the motor. This circuit, developed by CAV, represents a real improvement over the brutal spin-and-hit action of solenoid-operated and inertia clutches and should result in longer life for all components, including the flywheel ring gear.

Solenoids and relays can best be tested by bridging the large contacts. If the starter works, you know that the component has failed. Relays are sealed units and are not repairable. But most solenoids are at least amenable to inspection. Repairs to the series or shunt windings (illustrated in Figures 11-20 and 11-21) are out of the question unless the circuit has opened at the leads. Contacts can be burnished, and some designs have provision for reversing the copper switch element.

Starter drives

Most diesel starters feature positive engagement drives energized by the solenoid. The solenoid is usually mounted piggyback on the frame and the pinion moved by means of a pivoted yoke (Figure 11-20).

Regardless of mechanical differences between types, all starter drives have these functions:

- The pinion must be moved laterally on the shaft to engage the flywheel.
- The pinion must be allowed to disengage when flywheel rpm exceeds pinion rpm.
- The pinion must be retracted clear of the flywheel when the starter switch is opened.

Figure 11-22 illustrates a typical drive assembly. The pinion moves on a helical thread. Engagement is facilitated by a bevel on the pinion and the ring gear teeth of the flywheel. Extreme wear on either or both profiles will lock the pinion.

The clutch shown employs ramps and rollers. During the motor drive phase the rollers are wedged into the ramps (refer to Figure 11-23). When the engine catches, the rollers are freed and the clutch overruns. Other clutches employ balls or, in a few cases, ratchets. The spring retracts the drive when the solenoid is deactivated.

Inspect the pinion teeth for excessive wear, and chipping. Some battering is normal and does not affect starter operation. The clutch mechanism should be disassembled (if possible), cleaned, and inspected. Inspect the moving parts for wear or deformation, with particular attention to the ratchet teeth and the ramps. Lubricate with Aero Shell 6B or the equivalent. Sealed drives should be wiped with a solvent-wetted rag. Do not allow solvent to enter the mechanism, because it will dilute the lubricant and cause premature failure. Test the clutch for engagement in one direction of pinion rotation and for disengagement in the other.

11-22 Typical starter drive.
Marine Engine Div., Chrysler Corp.

11-23 Overrunning clutch.
Marine Engine Div., Chrysler Corp.

Adjustment of the *pinion throw* is important to ensure complete and full mesh at the flywheel. Throw is measured between the pinion and the stop ring as shown in Figure 11-24. Adjustment is by adding or subtracting shims at the solenoid housing (Figure 11-25), moving the solenoid mounting bolts in their elongated slots, or turning the yoke pin eccentric.

Charging systems

The charging system restores the energy depleted from the battery during cranking and provides power to operate lights and other accessories. It consists of two major components: an alternator and a regulator. The circuit can be monitored by an ammeter or a lamp and is usually fused to protect the generator windings.

Alternators generate a high frequency 3-phase AC voltage, which is rectified (i. e., converted to direct current) by internal diodes. Wrong polarity will destroy the diodes and can damage the wiring harness. Observe the polarity when installing a new battery or when using jumper cables. Before connecting a charger to the battery, disconnect the cables. Should the engine be started with the charger in the circuit the regulator might be damaged. Isolate the charging system before any arc welding is done. Do not disconnect the battery or any other wiring while the alternator is turning. And, finally, do not attempt to polarize an alternator. The exercise is fruitless and can destroy diodes.

11-24 Pinion clearance. Marine Engine Div., Chrysler Corp.

11-25 Shims between solenoid body and starter determine the pinion clearance. Marine Engine Div., Chrysler Corp.

Initial tests

The charge light should be on with the switch on and the engine stopped. Failure to light indicates an open connection in the bulb itself or in the associated wiring. Most charging-lamp circuits operate by a relay under the voltage regulator cover. Lucas systems employ a separate relay that responds to heat. The easiest way to check either type is to insert a 0–100A ammeter in series with the charging circuit. If the meter shows current and the relay does not close, one can safely assume that it has failed and should be replaced. The Lucas relay can be tested as shown in Figure 11-26. You will need a voltage divider and 2.2W lamp. Connect clip A to the 12V terminal. The lamp should come on. Leaving the 12V connection in place, connect clip B to the 6V tap. The bulb should burn for 5 seconds or so and go out. Move B to the 12V post and hold for no more than 10 seconds. Then move it to the 2V (single cell) tap. The bulb should come on within 5 seconds. These units do not have computerlike precision, and some variation can be expected between them. But the test results should roughly correlate with the test procedure. Do not attempt to repair a suspect relay.

11-26 Testing Lucas charging lamp relay. To distinguish these relays from turn-signal flashers, Lucas has coded them green. GM Bedford Diesel

Test the alternator output against the meter on the engine or by inserting a test meter in series between the B terminal and battery. Voltage is monitored with a meter in parallel with the charging circuit. Discharge the battery by switching on the lights and other accessories. Connect a rheostat or carbon pile across the battery for a controlled discharge. (Without this tool you will be reduced to guessing about alternator condition.) With the load set at zero, start the engine and operate at approximately midthrottle. Apply the load until the alternator produces its full rated output. If necessary open the throttle wider. An output 2–6A below rating often means an open diode. Ten amps or so below rating usually means a shorted diode. The alternator might give further evidence of a diode failure by whining like a wounded banshee.

The voltage should be 18–20V above the nominal battery voltage under normal service conditions. It might be higher by virtue of automatic temperature compensation in cold weather.

Assuming that the output is below specs, the next step is to isolate the alternator from the regulator. Disconnect the field (F or FD) terminal from the regulator and ground it to the block. Load the circuit with a carbon pile to limit the voltage output. Run the engine at idle. In this test we have dispensed with the regulator and are protecting the alternator windings with carbon pile. No appreciable output difference between this and the previous tests means that the regulator is doing its job. A large difference would indicate that the regulator was defective.

Late-production alternators often have integrated regulators built into the slip ring end of the unit. Most have a provision for segregating alternator output from the regulator so that "raw" outputs can be measured. The Delcotron features a shorting tab. A screwdriver is inserted into an access hole in the back of the housing (Figure 11-27); contact between the housing and the tab shorts the fields.

Note: The tab is within $3/4$ in. of the casting. Do not insert a screwdriver more than 1 in. into the casting.

Bench testing

Disconnect the battery and remove the alternator from the engine at the pivot and belt-tensioning bracket. Three typical alternators are shown in exploded view in Figures 11-28 through 11-30.

Remove the drive pulley. A special tool will be needed on some of the automotive derivations. Hold the fan with a screwdriver and turn the fan nut counterclockwise. Tap the sheave and fan off the shaft with a mallet. Remove the throughbolts and separate the end shields.

11-27 Location of Delcotron shorting tab.

11-28 US generic pattern alternator.

Inspect the brushes for wear. Some manufacturers thoughtfully provide a wear limit line on the brushes (Figure 11-31). Clean the holders with a non-petroleum-based solvent and check the brushes for ease of movement. File lightly if they appear to bind. The slip rings should be miked for wear and eccentricity. Ten to twelve thousandths should be considered the limit (Figure 11-32). Slip rings are usually, but not always, integral with the rotor. Removable rings are chiseled off and new ones pressed into place. Fixed rings can be restored to concentricity with light machining.

Determine the condition of the rotor insulation with a 120V test lamp (Figure 11-33). The slip rings and their associated windings should be insulated from the shaft and pole pieces. If you have access to an accurate, low-range ohmmeter, test for continuity between slip rings. The resistance might lead one to suspect a partial open; less could mean an intracoil short.

The stator consists of three distinct and independent windings whose outputs are 120° apart. It is possible for one winding to fail without noticeably affecting the others. Peak alternator output will, of course, be reduced by one third. Disconnect the three leads going to the stator windings. Many European machines have these leads soldered, while American designs generally have terminal lugs. When unsoldering, be extremely careful not to overheat the diodes. Exposure to more than 300°F will upset their crystalline structure. Test each winding for resistance. Connect a low-range ohmmeter between the natural lead and each of three winding leads as shown in Figure 11-34. Resistance will be quite low—on the order of 5 or 6 Ω—and becomes critical when one group of windings gives a different reading than the others.

Test the stator insulation with a 120V lamp connected as shown in Figure 11-35. There should be no continuity between the laminations and windings.

The next step is to check the diodes (Figure 11-36). You might already have had some evidence of diode trouble in the form of alternator whine or blackened varnish on the stator coils. The diodes must be tested with an ohmmeter or a test lamp of the same voltage as generator output.

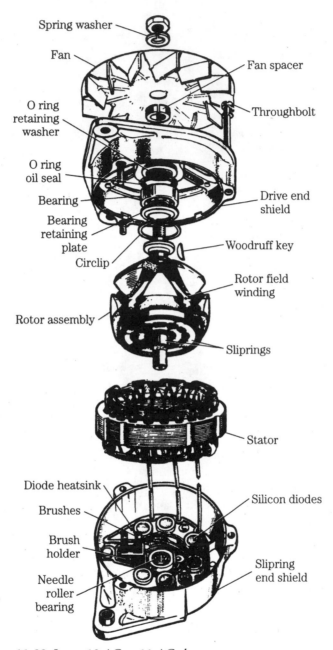

Spring washer

Fan

Fan spacer

O ring retaining washer

Throughbolt

O ring oil seal

Bearing

Drive end shield

Bearing retaining plate

Woodruff key

Circlip

Rotor field winding

Rotor assembly

Sliprings

Stator

Diode heatsink

Silicon diodes

Brushes

Brush holder

Slipring end shield

Needle roller bearing

11-29 Lucas 10-AC or 11-AC alternator. GM Bedford Diesel

Test each diode by connecting the test leads and then reversing their polarity. The lamp should light in one polarity and go out in the other. Failure to light at all means an open diode; continuous burning means the diode has shorted. In either case it must be replaced. You might use a low-voltage ohmmeter in lieu of a lamp.

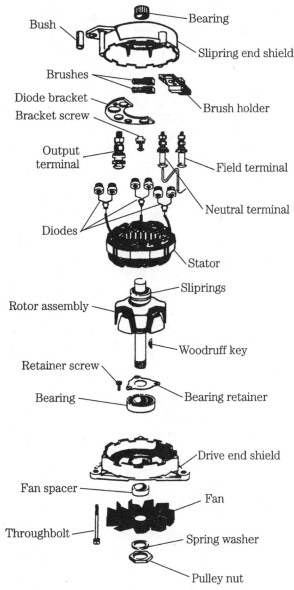

Bush

Bearing

Sliring end shield

Brushes

Diode bracket

Bracket screw

Brush holder

Output
terminal

Field terminal

Neutral terminal

Diodes

Stator

Sliprings

Rotor assembly

Woodruff key

Retainer screw

Bearing

Bearing retainer

Drive end shield

Fan spacer

Fan

Throughbolt

Spring washer

Pulley nut

11-30 Prestolite CAB-1235 or CAB-1245 alternator.
GM Bedford Diesel

Expect high (but not infinite) resistance with one connection, and low (but not zero) resistance when the two leads are reversed.

To simply service and limit the need for special tools, some manufacturers package mounting brackets with their diodes. The bracket is a heat sink and must be in intimate contact with the diode case. Other manufacturers take the more traditional approach and supply individual diodes, which must be pressed (not hammered) into

Wear limit line

11-31 Brush showing wear limit line. Chrysler Corp.

Dial gauge

Sliprings

11-32 Determining slipring concentricity. Marine Engine Div., Chrysler Corp.

15 W

110 V

11-33 Checking rotor insulation. GM Bedford Diesel

Ohms

11-34 Comparison test between stator windings. GM Bedford Diesel

11-35 Comparison test between stator windings. GM Bed Ford Diesel

their sinks. K-D Tools makes a complete line of diode removal and installation aids, including heat sink supports and diode arbors of various diameters. Figure 11-37 shows a typical installation with an unsupported heat sink. Other designs might require support.

Soldering the connections is very critical. Should the internal temperature reach 300°F the diode will be ruined. Use a 150W or smaller iron and place a thermal shunt between the soldered joint and the diode (Figure 11-38). The shunt might be in the form of a pair of needle-nosed pliers or copper alligator clips. In some instances there might not be room to shunt the heat load between the diode and joint (Figure 11-38) It is only necessary to twist the leads enough to hold them while the solder is liquid. Work quickly and use only enough solder to flow between the leads. More solder merely increases the thermal load and increases the chances that the diode will be ruined.

Alternator bearings are sealed needle and ball types. They are not to be disturbed unless noisy or rough. Then bearings are pressed off and new ones installed with the numbered end toward the arbor.

11-36 Diode testing. GM Bedford Diesel

11-37 Installing diode. Marine Engine Div., Chrysler Corp.

11-38 Two forms of heatsinks to protect the diode when soldering.
GM Bedford Diesel

Voltage regulation

Most alternator-based charging circuits employ voltage regulation (as opposed to voltage and current regulation). The regulator can be external to the alternator or integral with it. External regulators can be mechanical or solid state.

External regulators

Figure 11-39 is a schematic of a typical mechanical voltage regulator. The voltage-sensing winding is in parallel with the output and drives NC (normally closed) contacts. As the alternator comes up to speed, voltage increases until the winding develops a strong enough field to open the contacts. Output then passes through dropping resistor RF.

Depending on the make and model, voltage adjustment is accomplished by bending the stationary contact, moving the hinges in elongated mounting holes, or by a screw (Figure 11-40). In theory, the correct point gap should correspond with an output voltage of approximately 15V at 68°F or 28V for 24V systems. In practice, better results are had by measuring alternator output voltage at the battery terminals. Assuming that specifications are available, core and yoke gap adjustments also can be made.

Clean oxidized contacts with a riffle file or a diamond-faced abrasive strip. Do not use sandpaper or emery cloth. Inspect the dropping resistor (often found on the underside of the unit), springs, and contact tips for evidence of overheating. Check the regulator ground connection.

11-39 Voltage regulator in alternator circuit. Marine Engine Div., Chrysler Corp.

11-40 A typical relay and its adjustment points. Marine Engine Div., ChrysJer Corp.

Before discarding a defective regulator, attempt to discover why it failed. Burnt points or heat-discolored springs mean high resistance in the charging circuit or a bad regulator ground.

Solid-state regulators

Transistorized regulators are capable of exceedingly fine regulation, partially because there are no moving parts. Durability is exceptional. On the other hand any internal malfunction generally means that the unit must be replaced. As a rule, no repairs are possible. Failure can occur because of manufacturing error (this usually shows up in the first few hours of operation and is covered by warranty) high curent draws, and voltage spikes.

The mechanic must be particularly alert when working with transistorized circuits. The cautions that apply to alternator diodes apply with more force to regulators if only because regulators are more expensive. Do not introduce stray voltages, cross connections, reverse battery polarity, or open connections while the engine is running.

The regulator might be integral with the alternator or might be contained in a separate box. In general, no adjustment is possible; however, the Lucas 4TR has a voltage adjustment on its bottom, hidden under a dab of sealant.

Batteries

The battery has three functions: provide energy for the starting motor; stabilize voltages in the charging system; and, for limited periods, provide energy for the accessories in the event of charging-circuit failure. Because it is in a constant state of chemical activity and is affected by temperature changes, aging, humidity and current demands, the battery requires more attention than any other component in the electrical system.

Battery ratings

Starting a diesel engine puts a heavy drain on the battery, especially in cold climates. One should purchase the best quality and the largest capacity practical. The physical size of the battery is coded by its *group number*. The group has only an indirect bearing on electrical capacity but does ensure that replacements will fit the original brackets. In some instances a larger capacity battery might require going to another group number. Expect to modify the bracket and possibly to replace one or more cables.

The traditional measure of a battery's ability to do work is its *ampere-hour* (A-hr) *capacity*. The battery is discharged at a constant rate for 20 hr so that the potential of each cell drops to 1.75V. A battery that will deliver 6A over the 20-hr period is rated at 120 A-hr (6A × 20hr). You will find this rating stamped on replacement batteries or in the specifications.

Like all rating systems, the ampere-hour rating is best thought of as a yardstick for comparison between batteries. It has absolute validity only in terms of the orginally test. For example, a 120 A-hr battery will not deliver 120A for 1 hr, nor will it deliver 1200A for 6 minutes.

Cranking-power tests are more meaningful because they take into account the power loss that lead-acid batteries suffer in cold weather. At room temperature the battery develops its best power; power output falls off dramatically around 0°F. At the same time, the engine becomes progressively more difficult to crank and more reluctant to start. Several cranking-power tests are in use.

Zero cranking power is a hybrid measurement expressed in volts and minutes. The battery is chilled to 0°F; depending on battery size, a 150 or 300A load is applied. After 5 seconds the voltage is read for the first part of the rating. Discharge continues until the terminal voltage drops to 5V. The time in minutes between full charge and effective exhaustion is the second digit in the rating. The higher these two numbers are for batteries in the same load class, the better.

The *cold cranking performance* rating is determined by lowering the battery temperature to 0°F (or, in some instances, 20°F) and discharging for 30 seconds at such a rate that the voltage drops below 1.2V per cell. This is the most accepted of all cold weather ratings and has become standard in specification sheets.

Battery tests

As the battery discharges, some of the sulfuric acid in the electrolyte decomposes into water. The strength of the electrolyte in the individual cells is a reliable index of the state of charge. There are several ways to determine acidity, but long ago technicians fixed on the measurement of specific gravity as the simplest and most reliable.

The instrument used is called a *hydrometer*. It consists of a rubber bulb, a barrel, and a float with a graduated tang. The graduations are in terms of specific gravity. Water is assigned a specific gravity of 1. Pure sulfuric acid is 1.83 times heavier than water and thus has a specific gravity of 1.83. The height of the float tang above the liquid level is a function of fluid density, or specific gravity. The battery is said to be fully charged when the specific gravity is between 1.250 and 1.280.

An accurate hydrometer test takes some doing. The battery should be tested prior to starting and after the engine has run on its normal cycle. For example, if the engine is shut down overnight, the test should be made in the morning, before the first start. Water should be added several operating days before the test to ensure good mixing. Otherwise the readings can be deceptively low.

Use a hydrometer reserved for battery testing. Specifically, do not use one that has been used as an antifreeze tester. Trace quantities of ethylene glycol will shorten the battery's life.

Place the hydrometer tip above the plates. Contact with them can distort the plates enough to short the cell. Draw in a generous supply of electrolyte and hold the hydrometer vertically. You might have to tap the side of the barrel with your finger-nail to jar the float loose. Holding the hydrometer at eye level, take a reading across the fluid level. Do not be misled by the meniscus (concave surface, Figure 11-41) of the fluid.

American hydrometers are calibrated to be accurate at 80°F. For each 10°F above 80°F, add 4 points (0.004) to the reading; conversely, for each 10°F below the standard, subtract 4 points. The standard temperature for European and Japanese hydrometers is 20°C, or 68°F. For each 10°C increase add 7 points (0.007); subtract a like amount for each 10°C decrease. The more elaborate hydrometers have a built-in thermometer and correction scale.

All cells should read within 50 points (0.050) of each other. Greater variation is a sign of abnormality and might be grounds for discarding the battery. The relationship between specific gravity and state of charge is shown in Figure 11-42.

The hydrometer test is important, but by no means definitive. The state of charge is only indirectly related to the actual output of the battery. Chemically the battery might have full potential, but unless this potential passes through the straps and terminals, it is of little use.

Perhaps the single most reliable test is to load the battery with a rheostat or carbon pile while monitoring the terminal voltage. The battery should be brought up to full charge before the test. The current draw should be adjusted to equal three times the

Float

Glass
tube

11-41 Reading hydrometer.
Marine Engine Div., Chrysler Corp.

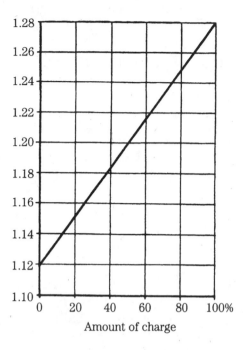

11-42 Relationship between state of charge and specific gravity.
Marine Engine Div., Chrysler Corp.

ampere-hour rating. Thus, a 120 A-hr battery would be discharged at a rate of 360A. Continue the test for 15 seconds and observe the terminal voltage. At no time should the voltage drop below 9.5V.

In this test, sometimes called the *battery capacity* test, we used voltage as the telltale. But without a load, terminal voltage is meaningless. The voltage remains almost constant from full charge to exhaustion.

Battery maintenance

The first order of business is to keep the electrolyte level above the plates and well into the reserve space below the filler cap recesses. Use distilled water. Tap water might be harmful, particularly if it has iron in it.

Inspect the case for cracks and acid seepage. Periodically remove the cable clamps and scrape them and the battery terminals. Look closely at the bond between the cables and clamp. The best and most reliable cables have forged clamps, solder-dipped for conductivity. Replace spring clip and other clever designs with standard bolt-up clamps sweated to the cable ends. After scraping and tightening, coat the terminals and clamps with grease to provide some protection from oxidation.

The battery case should be wiped clean with a damp rag. Dirt, spilled battery acid, and water are conductive and promote self-discharge. Accumulated deposits can be cleaned and neutralized with a solution of baking soda, water, and detergent. Do not allow any of the solution to enter the cells, where it would dilute the electrolyte. Rinse with clear water and wipe dry.

Charging

Any type of battery charger can be used—selenium rectifier, tungar rectifier, or, reaching way back, mercury arc rectifier. Current and voltage should be monitored and there should be a provision for control. When charging multiple batteries from a single output, connect the batteries in series as shown in Figure 11-43.

Batteries give off hydrogen gas, particularly as they approach full charge. When mixed with oxygen, hydrogen is explosive. Observe these safety precautions:

- Remove all filler caps (to prevent pressure rise should the caps be clogged).
- Charge in a well-ventilated place remove from open flames or heat.
- Connect the charger leads before turning the machine on. Switch the machine off before disconnecting the leads.

In no case should the electrolyte temperature be allowed to exceed 115°F. If your charger does not have a thermostatic control, keep track of the temperature with an ordinary thermometer.

Batteries can be charged by any of three methods. *Constant-current* charging is by far the most popular. The charging current is limited to one-tenth of the ampere-hour rating of the battery. Thus a 120 A-hr battery would be charged at 12A. Specific gravity and no-load terminal voltage are checked at 30-minute intervals. The battery might be said to be fully charged when both values peak (specific gravity 1.127–1.129, voltage 15–16.2V) and hold constant for three cranking intervals.

A *quick charge*, also known as a booster or hotshot, can bring a battery back to life in a few minutes. The procedure is not recommended in any situation short of an emergency, because the high-power boost will raise the electrolyte temperature and might cause the plates to buckle. Disconnect the battery cables to isolate the generator or alternator if such a charge is given to the battery while it is in place.

A constant-voltage charge can be thought of as a compromise between the hotshot and the leisurely constant-current charge. The idea is to apply a charge by keeping charger voltage 2.2–2.4 V higher than terminal voltage. Initially the rate of charge is quite high; it tapers off as the battery approaches capacity.

11-43 When charging multiple batteries, connect in series. Marine Engine Div., Chrysler Corp.

Battery hookups

One of the most frequently undertaken field modifications is the use of additional batteries. The additional capacity makes starts easier in extremely cold weather and adds reliability to the system.

To increase capacity add one or more batteries in parallel—negative post connected to negative post, positive post to positive post. The voltage will not be affected, but the capacity will be the sum of all parallel batteries.

Connecting in series—negative to positive, positive to negative—adds voltage without changing capacity. Two 6V batteries can be connected in series to make a physically large 12V battery.

Cable size is critical because the length and cross-sectional area determine resistance. Engineers at International Harvester have developed the following recommendations, which can be applied to most small, high-speed diesel engines:

- Use cables with integral terminal lugs.
- Use only rosin or other noncorrosive-flux solder.
- Terminal lugs must be stacked squarely on the terminals. Haphazard stacking of lugs should be avoided.
- Where the engine frame is used as a ground return, it must be measured and this distance added to the cable length to determine the total length of the system. Each point of connection with the frame must be scraped clean and tinned with solder. There should be no point of resistance in the frame such as a riveted joint. Such joints should be bridged with a heavy copper strap.
- Pay particular attention to engine-frame grounds. If the engine is mounted in rubber it is, of course, electrically isolated from the frame.
- Check the resistance of the total circuit by the voltage drop method or by means of Ohm's law ($R = E/I$).
- Use this table as an approximate guide to cable size for standard-duty cranking motors.

System voltage	Maximum resistance	Cable size & length
12V	0.0012Ω	Less than 105 in., No. 0
		105 to 132 in., No. 00
		132 to 162 in., No. 000
		162 to 212 in., No. 0000 in parallel.
24V	0.0020Ω	Less than 188 in., No. 0
		188 to 237 in., No. 00
		237 to 300 in., No. 0000
		300 to 380 in., No. 0000, or two No. 0 in parallel.

- Suggested battery capacity varies with system voltage (the higher the voltage, the less capacity needed for any given application), engine displacement,

	(1) Simple addition of battery	(2) Switch used	(3) Charging distributor used
Wiring	Alternator — Main battery — Aux. battery	Alternator — Switch — Main battery — Aux. battery	Alternator — Charging distributor — Main battery — Aux. battery
Alternator (operating)	Main battery / Aux. battery	Main battery / Aux. battery	Main battery / Aux. battery
Alternator (off) (Aux. battery operating)			
Starter started			
Remarks	Main battery discharge due to use of aux. battery	Main battery discharge by turning on the switch for charging aux. battery.	Discharge from the main battery prevented.

11-44 Hard wiring batteries in parallel divides the load equally between both. Installing a manual switch in the B+ line confines the load to the auxiliary battery *until the switch is closed for charging*. When this happens, the main battery promptly discharges into the auxiliary. The best solution is to purchase a charging distributor. To avoid problems, furnish the vendor with a complete schematic and alternator characteristics.

compression ratio, ambient air temperature, and degree of exposure. Generalizations are difficult to make, but typically a 300 CID engine with a 12V standard-duty starter requires a 700A battery for winter operation. This figure is based on SAE J-5371 specifications and refers to the 30-second output of a chilled battery. At 0°F capacity should be increased to 900A. The International Harvester 414 CID engine requires 1150, or 1400A at 0°F. Battery capacity needs roughly parallel engine displacement figures, with some flattening of the curve for the larger and easier-to-start units. In extremely cold weather—below −10°F—capacity should be increased by 50% or the batteries heated.

• RV and boat owners often install a second battery to support accessory loads. Arranging matters so that there is always power available for starting requires special hardware; otherwise, the state of charge of either battery is the average of the two. Figure 11-44 illustrates this proposition.

12

Cooling systems

Roughly 30% of the heat energy produced by combustion must be dissipated by the cooling system. In addition, the cooling system can be required to absorb heat from aftercoolers, engine and transmission oil coolers, hydraulic oil coolers, and other sources.

Air cooling

The great advantage of air cooling is its simplicity. There are no radiators, pumps, or hoses to add dead weight and eventually fail. On the other hand, the cooling fins and aluminum castings that promote heat transfer also transmit sound. Air-cooled engines are noisy. Nor does air cooling provide the precise temperature control necessary for good efficiency and low exhaust emissions. The centrifugal fan behaves like a turbocharger, pumping out too much air at high speeds and too little when the engine bogs under load.

Most air-cooled engines are small single- and twin-cylinder units developing less than 40 hp. But the concept is also applied to larger engines. Deutz builds a range of modular air-cooled engines of up to 500 hp, some of which are even used in marine applications. Most of the world's armored vehicles are powered by air-cooled diesels.

Figure 12-1 shows the cooling arrangement for small utility engines. The flywheel-mounted fan generates air flow that is directed over the cylinder by means of shrouding. Few of these engines are equipped with temperature gauges, which is a serious oversight. The operator can however measure oil temperature with a thermometer placed in the dipstick boss to a depth of about 5/16 in. below the end of the stick, but well clear of metal surfaces. Maximum permissible oil temperature is a judgment call, with one manufacturer suggesting that it should not exceed 210°F (99°C) above ambient.

The thermostatically controlled shutter used by Onan gives an idea of the normal range of air-outlet temperatures (Figure 12-2). The thermostat, mounted in the air outlet, begins to open the shutter when temperature reaches 120°F (49°C) and

12-1 Air-flow for an air-cooled engine. Courtesy Onan

extends fully at 140°F (60°C). A second thermostat shuts off fuel delivery when outlet-air temperature reaches 250°F (121°C). It is fair to say that outlet-air temperature should not exceed 45°F above ambient under load.

Chronic overheating means that the engine is undersized for the application. Other than periodically cleaning the fins and checking the fit of the tin work, there is little a mechanic can do to improve cooling. A very light coat of dull black paint applied over bare metal on exposed surfaces makes a marginal improvement in radiation.

12-2 Thermostatically controlled shutter provides improved temperature control. Onan

Liquid cooling

Early engines were cooled by water in a hopper above the cylinders that was refilled as the water boiled off. Modern practice is to employ closed systems with a radiator or other form of heat exchanger, one or more circulation pumps, and a thermostat. Figure 12-3 illustrates the basic system. When the engine is cold, the thermostat closes to confine most of the coolant within the water jacket. A small fraction of the coolant makes its way to the radiator through an internal bleed port

12-3 Coolant flow in system without a provision for air purging. Thermostat closed (A) and open (B). International

in the thermostat. As shown in the drawings, the bypass often takes the form of a small-diameter hose running from below the thermostat housing to the pump. Some circulation is necessary to prevent local hot-spots in the water jacket.

As water temperature rises—a typical figure is 190°F (87.8°C)—the thermostat opens. Flow then passes from the water jacket through the top hose to the radiator and out of the lower hose to the water pump.

Most cooling systems include one or two additional hoses that function to vent air to the radiator or expansion tank. The system depicted in Figure 12-4 employs two lines that convey aerated coolant to the radiator header. Air enters from splash

12-4 Coolant flow in a self-purging system. Thermostat closed (A) and open (B). International

12-5 Marine application using a seawater-cooled heat exchanger and expansion tank. Caterpillar Inc.

entrapment (as, for example, when the radiator is filled from a bucket), past the water-pump seal, and by way of compression leaks across the head gasket. Besides reducing heat transfer—air removes heat 3500 times less efficiently than water—air increases the tendency of the water pump to lose pressure through cavitation.

Modern cooling systems employ an expansion tank, connected to the radiator-cap overflow and discharging into the radiator-return line. This expansion, surge, or degassing tank vents entrapped air and exhaust gases, collects overflow, and provides convenient means of replenishing the coolant. In order to vent gases, the tank must be located above the thermostat housing, which is normally at the highest wetted point on the engine block.

Figures 12-5 and 12-6 illustrate two fresh-water marine systems, one using a seawater-cooled heat exchanger and the other a keel cooler. Expansion tanks are clearly shown.

As illustrated in Figure 12-7, standard practice is to distribute coolant to the cylinder liners through an external manifold. Many engines have an oil cooler in series with the main cooling circuit, and multiple liquid-cooled accessories, such as an aftercooler, turbocharger, torque converter, cab heater, and vehicle brakes, plumbed in parallel.

Coolant

Diesel manufacturers recommend a 50-50 mixture of low-silicate ethylene glycol and distilled water. This mixture gives −34°F (−37°C) protection against freezing and raises the boiling point to 226°F (107°C). Pressurization raises the boiling point further.

1. turbocharger
2. aftercooler, jacket water cooled
3. jacket water inlet connection
4. jacket water inlet connection
5. expansion tank
6. jacket water pump
7. keel cooler
8. bypass filter
9. duplex full-flow strainer
10. shut-off valve
11. auxiliary expansion tank
12. flexible connection

12-6 Marine application with a keel cooler. Courtesy Caterpillar Inc.

Some diesel manufacturers permit the use of standard automotive antifreezes that meet ASTM D3306 specifications. Others insist upon the more stringent D4985 or D6210 specification. D4985 has an inhibitor life of one year, and D6210 is good for 3000 operating hours or three years. The additive package in Caterpillar ELC (Extended Life Coolant), which is incompatible with other antifreezes, provides protection for 12,000 hours or six years. Depending upon the type of antifreeze used, long-term protection may require that additive package be periodically replenished. Check with your dealer.

Soluble oils and methyl alcohol have no place in modern engines.

While there is no substitute for distilled or ionized water, there are times when operators are forced to use whatever is available. The water should meet these minimum specifications:

Cylinder Head **Water Bypass** **Thermostat Housing** **Water Pump** **Radiator** **Cylinder Liners** **Oil Cooler** **Cylinder Block**

12-7 A radiator, oil cooler, and water jacket make up the basic coolant circuit. Caterpillar Inc.

Chloride (Cl)	40 mg/L (2.4 grains/US gal)
Sulfate (SO$_4$)	100 mg/L (5.9 grains/US gal)
Total hardness	170 mg/L (10.0 grains/US gal)
Total solids	340 mg/L (20.0 grains/US gal)
Acidity	pH 5.5–9.0

The local water utility or agricultural agent can provide data on water quality.

CAUTION: do not overfill radiators. Topping off the radiator or expansion tank when the engine is cold merely wastes coolant, since the surplus goes out the overflow when the engine warms. Chronic overfilling dilutes the antifreeze.

A major drawback associated with ethylene glycol is the way it reacts with lube oil. Should a leak develop at the head gasket or oil cooler, shut the engine down immediately and make the necessary repairs. Assuming that the engine will still turn over, you might be able to avert a complete teardown by flushing the lubrication system with Butyl Cellosolve or an equivalent product. But check with the factory first, since the flushing procedure was developed for Detroit two-cycles and could damage more heavily loaded engines.

The procedure consists of replacing the contaminated lube oil with a 50-50 mix of Cellosolve and SAE 10 oil. The engine is run at 1200 no-load rpm for one hour, then for another 15 minutes with pure SAE 10 in the crankcase. If the bearings survive, that is, if oil pressure remains steady and no knocks are heard, the operation was successful and the engine may be returned to service.

Overheating

Most diesel temperature gauges are calibrated to show overheating at around 226°F (108°C), but panel gauges cannot be trusted. Verify with an accurate thermometer, and preferably one of the remote-sensing infrared types that can be used to detect hot spots in the radiator. Other signs of possible overheating include steam from the radiator overflow after shutdown, coolant leaks, and low coolant levels. Low coolant levels can be both a cause of overheating and its result, as coolant boils off and escapes out the vented cap or expansion tank. In any event, refill the radiator and see if the problem recurs.

Severe overheating results in blown head gaskets, warped or cracked cylinder heads, and scored pistons. The pistons for the aft cylinders usually fail first. As a point of interest, the style of combustion chamber influences how pistons seize. Pistons for IDI engines generally bind at the area just below the upper ring land. DI pistons are more likely to seize on their thrust faces.

Table 12-1 lists the most common causes of overheating.

Table 12-1. Overheating problems and causes

Problem	Possible causes
Low coolant level	External leaks at hoses, radiator, and radiator cap
	Internal leaks at head gasket, cylinder head, aftercooler, torque-converter cooler, etc.
Low heat transfer	Clogged radiator
	Scale accumulations in water jacket
	Marine growth on keel cooler or insufficient flow of raw water through heat exchanger
Insufficient coolant flow	Thermostat stuck closed
	Loose water-pump belts
	Water-pump failure
	Clogged radiator core
Insufficient coolant pressure	Failed radiator pressure cap
	Failed coolant-pressure relief valve (when fitted)
Coolant overflow	Combustion gases entering system—loose or cracked cylinder head, leaks at head gasket, precombustion chamber, cylinder liner
	Entrapped air in cooling system
Insufficient air flow through radiator	Low or no fan speed—electric or hydraulic fan drive failure, loose belts, worn pulleys
	Fan installed backwards
	Shutter not opening
	Shrouding not installed properly
High inlet air temperature or restriction	Clogged air cleaner
	Clogged aftercooler
	Turbocharger failure
Exhaust restriction	Turbocharger failure
	Water in muffler or loose muffler baffle
	Clogged particulate trap/catalytic converter (when fitted)

Coolant leaks

Loss of coolant results from intrusion of combustion gases into the water jacket, a failed radiator cap or cap seal, and leaks. Most leaks are obvious, but look for telltale rust stains at gasketed joints with special attention to the block/head interface. Freeze plugs rust from the inside out. Very small leaks can be found by coloring the coolant with dye and pressurizing the system with a hand pump.

Corrosion and scale

A fairly accurate idea of the corrosion present in the water jacket can be had by removing the radiator hose from the thermostat housing. A film of light rust that wipes off with rag may be present on new engines and has no significance. Layered rust is cause for concern.

Most corrosion can be blamed on poor maintenance. Ethylene glycol antifreeze is safe so long as its additive package remains active. When in doubt about the condition of the antifreeze, change it. Antifreeze test kits for nitrite and phosphate inhibitors quickly pay for themselves in fleet service.

The pH level of the coolant should be neutral, that is between 7.0 and 10.5. Below 7.0 the coolant is acidic and dissolves iron; above 10.5 it turns alkaline and attacks aluminum, solder, and other nonferrous metals. The test apparatus is available from chemical supply houses.

Another, and sometimes difficult to correct, source of corrosion is electrolysis. When two dissimilar metals are subject to voltage, the least "noble" of these metals erodes. The coolant acts as the conductor and aluminum becomes the sacrificial metal, eroding about twice as rapidly as cast iron under the same voltage. Check for loose, dirty, or rusted ground connections, missing engine-to-ground straps, and for the presence of sacrificial anodes on marine applications. Using a digital ohmmeter, check the resistance between each grounded electrical component and the battery negative post. Resistance should be less than 0.03Ω. A higher reading means that the component is not properly grounded.

Exhaust gas seeping into the coolant also contributes to corrosion. Make up a hose connection to a sensor boss on the thermostat housing and collect the coolant in a glass container. Run the engine at normal temperature under load. An occasional bubble is acceptable, but a steady stream of bubbles indicates an exhaust leak.

Scale consists of calcium carbonate and sulfate, iron, copper, silica, and trace metals. According to General Motors, 1/16 in. of scale is the thermal equivalent of $4\frac{1}{2}$ in. of cast iron. The primary source of scale is hard water that does not meet the minimum specifications listed earlier.

Cleaning

Reverse flushing removes loose rust and silt from the radiator and block. To flush the radiator, disconnect both hoses and insert a flushing gun, connected to a 100-psi air line, into the lower hose (Figure 12-8A). Fill the radiator and inject air in short bursts. Continue until the water flows freely. With the thermostat removed, do the same for the block (Figure 12-8B).

Discharge hose

Flushing gun

A

Discharge hose
connected to
water pump

B

12-8 Reverse flushing the radiator
(A) and the block (B). GM Bedford Diesel

Two-part cleaners, consisting of oxalic acid and a neutralizing compound, can remove light rust and scale deposits without taking the engine out of service. Any of the commercial products work, or you can mix your own acid by adding 2 lb of sodium bisulfate with 10 US gal of water. The neutralizer consists of ½ lb of sodium carbonate crystals per 10 gal of water.

The presence of oil in the coolant can usually be traced to a leaking oil or transmission cooler, although other sources cannot be overlooked. A leak can develop across any interface that separates pressurized lube oil from coolant. Oil leaks result in local overheating that does not register on the temperature gauge. According to Detroit Diesel 1.25% of oil by volume in the coolant increases firedeck temperature by 15%. Once the leak is found and repaired, drain the cooling system and refill with water and two cups of non-foaming dishwasher detergent. Run the engine for 20 minutes or so, adding more detergent as necessary to emulsify the oil. Drain the system, flush with water to remove all traces of detergent, and refill with approved coolant.

Cavitation erosion

Wet cylinder liners flex under combustion pressures and pull away from the surrounding coolant. Air bubbles form in the void, attach themselves to the cylinder liners and implode, leaving tiny pits in their wake. The action is progressive and, at some point, the pitting breaks through to admit coolant into the cylinder (Figure 12-9).

New liners have protection in the form of an oxide coating. But the oxide eventually pits and one must rely on additives, such as nitrates, to inhibit bubble formation and slow the rate of metal loss. The operator should avoid lugging the engine, which increases the amplitude of vibration, and do what he can to minimize piston slap. The fewer cold starts, the better. But cavitation erosion will be with us as long as diesel engines use wet liners.

Overcooling

Standard practice is to size the cooling system to yield a top radiator-tank temperature of 210°F (99°C) at a maximum ambient temperature of 110°F (43°C). A properly calibrated thermostat and the use of radiator shutters in cold climates should prevent overcooling and the resulting gum and carbon deposits.

12-9 Cavitation erosion results in coolant leaks into the cylinders. Caterpillar Inc.

12-10 Conventional downflow radiator. The baffle distributes coolant over the width of the core and traps gas bubbles that vent through the bleed tube. International

Radiators

Heavy-duty radiators have large frontal areas and free-flowing cores with no more than 10 or 11 fins/in. These radiators are square, or nearly square, so that a single fan can pull or push air over most of the core surface. As shown in Figure 12-10, coolant flows down from the header tank, usually made of brass, across a baffle, and through the vertical copper tubes. In the better examples, the tubes are brazed to the header tank. The Modine Beta Weld process, originally developed for off-road-vehicle radiators, produces a stronger and more reliable joint than soldering.

Radiators for industrial/commercial applications have a built-in safety factor known as the drawdown capability. Standards vary somewhat with the engine maker, but a properly sized system can lose between 10% or 15% of its coolant without overheating. Figure 12-11 charts this relationship.

Passenger-car radiators come out of a different tradition. Low hood lines result in elongated radiators, which often require two fans to cool. Since the early 1990s, aluminum has been specified for the core, and header tanks are often made of plastic. Designers compensate for loss of heat transfer—aluminum conducts heat only about half as well as copper—by narrowing the fin spacing and making the cores thicker. These almost solid radiators require large amounts of fan power to cool.

Automotive cooling systems must be accepted as they are, since the engineering that goes into them and underhood space limitations preclude much by way of modification. But cooling systems for stationary applications can often be improved with a bit of judicious tinkering. For example, most have simple box shrouds that can be replaced by much more efficient venturi shrouds. Other modifications are discussed below under "Fans."

As far as routine maintenance goes, green slime (chromium hydroxide) and sediment can be removed by reverse flushing the core or with oxalic acid. When overheating persists, farm out the radiator to a specialist for chemical cleaning.

12-11 The drawdown rating refers to the system's ability to function with reduced coolant volume. Acceptable drawdown capacities are shown here for systems with capacities of up to 360 quarts. International

Fins, especially closely spaced fins, tend to clog and should periodically be cleaned with compressed air, high-pressure water, or steam. Some of the worst offenders are off-road vehicles that splatter their radiators with mud and fork-lift trucks that handle cotton or other fibrous products.

Leaks in brass and copper radiators can often be repaired on site. The secret lies in the soldering, or more exactly, in the preparation for soldering. Remove all traces of grease and oil from the damaged area. Then, using fine emery paper, sand down the area to bright metal. Apply generous amount of flux (a mixture of muriatic acid and zinc powder) and solder. Hard, 60/40 solder makes a stronger joint than the softer grades. If the solder bubbles and skates, the surface is still not clean. The solder should sink into the base metal and harden with a mirror-like glaze. Once the repair is made, wipe off all traces of flux.

Pressure caps

Pressure, generated by the radiator cap, raises the boiling point of the coolant (Figure 12-12) and reduces the tendency of the water pump to cavitate. The cap includes two check valves, one that opens under vacuum (1) and a second, spring-loaded valve (2) that regulates system pressure (Figure 12-13). Pressures range from 3 or 3 psi to more than 15 psi.

A pressure cap that fails to hold pressure can cause boiling and loss of coolant, which ultimately results in overheating. A cap that fails to vent after shutdown, when the volume of coolant shrinks, often collapses the radiator hoses.

And finally, a word of caution. Radiator caps cam open in two stages. The first stage releases pressure and diverts coolant downward, away from the mechanic's hands. The second stage frees the cap from the radiator neck. All manufacturers warn

12-12 Graph showing how the boiling point of water varies with altitude and pressure. Caterpillar Inc.

that the engine should be allowed to cool before removing the cap. Risk increases with temperature and altitude. Releasing pressure on an overheated engine releases a geyser of boiling coolant and superheated steam that, while directed downward, rebounds off adjacent surfaces.

Fans

Fans for stationary applications operate continuously with power transmitted by one or more v-belts from the engine crankshaft. At 35 mph or so, road-going vehicles generate enough ram air velocity to dispense with the fan. Most passenger cars and light trucks use electric fans that cycle on and off in response to radiator header-tank temperature. Some earlier vehicles employed belt-driven fans with viscous clutches

1 Vacuum valve
2 Pressure valve
3 Filler cap scal
4 Overflow pipe

12-13 Cutaway view of a pressure cap. GM Bedford Diesel

Table 12-2. Recommended v-belt tension*

Belt width	Tension	
	New	Used
3/8	445 +/− 22N (100 +/− 5 lb)	400 +/− 22N (90 +/− 10 lb)
$^1/_2$	534 +/− 22N (120 +/− 5 lb)	400 +/− 44N (90 +/− 10 lb)
5V	534 +/− 22N (120 +/− 5 lb)	400 +/− 44N (90 +/− 10 lb)
11/16	534 +/− 22N (120 +/− 5 lb)	400 +/− 44N (90 +/− 10 lb)
$^3/_4$	534 +/− 22N (120 +/− 5 lb)	400 +/− 44N (90 +/− 10 lb)
15/16	534 +/− 22N (120 +/− 5 lb)	400 +/− 44N (90 +/− 10 lb)
6PK	667 +/− 22N (120 +/− 5 lb)	467 +/− 44N (105 +/− 10 lb)
8K	800 +/− 22N (120 +/− 5 lb)	489 +/− 44N (110 +/− 10 lb)

*Courtesy Caterpillar Inc.

that used silicone as the working fluid. Clutch action was less than positive, allowing the fan to turn at 700–1000 rpm when it should have been disengaged. And when fully engaged, the clutch slipped, reducing fan speed by about 5%.

Belt-driven on/off fans represent more recent thinking. Fans for light trucks employ electromagnetic clutches, similar to those used on air-conditioning compressors. Pneumatic clutches are favored for larger trucks with air brakes. The engine control unit (ECU) controls fan's on-off time in response to coolant temperature and other variables.

The primary maintenance requirement for belt-driven fans is to periodically check belt tension with a Borroughs or equivalent gauge (Table 12-2). You can get a rough idea of tension by applying thumb pressure to the longest belt run between pulleys. A half-inch of "give" is about right, but err on the loose side since belts are less expensive than bearings. Examine belts for oil damage, heat checking, and wear. Severe belt or pulley wear causes the belts to ride on the bottom of the pulley grooves. Paired belts should be replaced as a set—the stretch of the worn belt cannot be adjusted for without over-tensioning the replacement.

Check bearings by removing the belts and turning driven components by hand. "Hard spots" or perceptible side play means that the associated bearing should be replaced. Viscous-clutch units sometimes fail to reach speed as they age. The only way this fault can be detected is to measure airflow velocity with a Caterpillar 8T-2700 or equivalent tool.

There is little that can be done to improve the performance of clutched fans, which have been, or should have been, engineered for the application. But fixed-speed fans can often benefit for creative tinkering. Modifications include:

- Removing restrictions to air flow in front of the radiator and aft of the fan.
- Substituting a venturi-type shroud for the box shroud fitted to many industrial engines. Blade tips should come within 0.5 in. or less of the shroud.
- Increasing the radiator area swept by the fan. This can be done by specifying a larger diameter fan and, on suction fans, by displacing the fan at least on blade width behind the radiator.

- Investing in a high-efficiency fan and adjusting the drive ratio to turn it at the recommended blade-tip velocity.

Thermostats

A thermostat is a heat-sensitive valve that opens to admit coolant to the radiator when the temperature of coolant in the cylinder head reaches a predetermined temperature. Most thermostats open at 190°F (87.8°C).

Figure 12-14 shows a wax-pellet thermostat in the closed and open positions. When closed, a small amount of coolant passes through a bleed port to the radiator, but most returns to the pump through the bypass valve (8 in the drawing). As the coolant warms, the wax pellet (6) liquifies and expands to open the valve (2) to the radiator. Action is progressive. In the partially open position, coolant flow splits between the bypass valve and the radiator. At full open, all flow goes to the radiator. As the engine warms, radiator header-tank and cylinder temperatures should equalize.

Brass frames tend to break, and the metal bellows (rather than the telescoping tubes shown in the drawing) develop cracks. To check temperature response, heat the thermostat in water, supporting it away from the metal sides of the container (Figure 12-15). The temperature rating, almost always stamped on the frame, refers to the temperature at which the thermostat cracks open. Raise the water temperature to about 25°F (15°C) above the temperature rating, adding antifreeze as necessary to prevent boiling. Ten minutes should be enough time for the unit to open fully.

Water pumps

As shown in Figure 12-16, the water pump consists of a cast-iron or aluminum body, bearing assembly, ceramic face seal, shaft, and pressed-on impeller. Light duty pumps employ prelubricated bearings; the better types are plumbed into the engine

1 Shaft
2 Valve
3 Sheet
4 Frame
5 Spring
6 Pellet
7 Spring
8 Bypass valve
9 Snapring

12-14 Pill-type thermostat. Marine Engine Div. Chrysler Corp.

12-15 Testing a thermostat. International

1 Lockwire
2 Snapring
3 Pulley hub
4 Bearing
5 Water seal
6 Ceramic seat
7 Housing
8 Machine screw
9 Plate
10 Gasket
11 Shaft
12 Impeller

12-16 Typical water pump. Marine Engine Div., Chrysler

oiling circuit. All have a bleed port outboard of the seal on the underside of the body casting. The port must remain open to provide an escape route for coolant that gets past seal faces and, on pressure-lubricated pumps, to prevent coolant contamination of the lube oil.

Pump failure is signaled by:

- Leaks from the bleed port. Some moisture is normal, but coolant streaks in this area mean seal and possible bearing failure.
- Perceptible side play or "hard spots" as the shaft is turned by hand. In severe cases, bearing wear allows the impeller to make rubbing contact with the pump body.
- Low pressure. Some engines have test ports on the inlet and outlet sides of the pump so that the pressure rise—usually in the order of 12 psi—can be measured. In the absence of these ports, about all one can do is intuit pump performance from the volume of coolant going into the radiator.

Water pumps for non-passenger car applications have factory support in the form of replacement seals and rebuild kits (Figure 12-17). Most pumps require the use of an arbor press to remove and install the impeller and bearing pack.

Raw-water pumps assist the engine water pump in marine applications and in applications with remotely mounted radiators. The Jabsco unit in Figure 12-18 employs prelubricated bearings and a mechanical face seal. The symmetrical impeller enables the pump to be driven in either direction by the expedient of swapping inlet and outlet port connections. Some models include a drain cock; for the unit illustrated, cover screws must be loosened in preparation for freezing temperatures.

To disassemble, remove the cover-retaining screws, end cover (2), and impeller (5). If the pump has much time on it, the impeller will be stuck. Remove it with a suitable puller or by spreading plier jaws under the impeller and tapping the shaft with a brass drift. Remove the cam screw and the cam (6). Replace the wear plate (7) as a precautionary measure.

Support the pump body in a vise and loosen the pinch bolt (24) that secures the body to the bearing housing (15). Remove the slinger (14) and pry off the inner seal

12-17 Perkins rebuild kit. Most manufactures can supply the seal separately.

1 Screw, cover retaining
2 Pump, end cover
3 Gasket, end cover
4 Spline seal
5 Impeller
6 Cam
7 Wearplate
8 Retaining ring
9 Seal
10 Seal seat
11 Seat cup rubber
12 Cam screw
13 Body, water pump

14 Slinger
15 Bearing housing
16 Seal
17 Bearing
18 Spacer
19 Retaining ring bearing
20 Seal
21 Retaining ring, bearing
22 Nut
23 Washer, spring lock
24 Bolt, hexagon head
25 Shaft

12-18 Jabsco raw-water pump. GM Bedford Diesel

from the pump body casting. Press the shaft (25) and bearings out of the casting. Working from the impeller side, drive out the inner seal. All that remains is to separate the bearing pack from the shaft by removing the circlip (8).

Hoses

Hose condition is best checked by feel. Replace hoses that feel soft after the engine has shut down and pressure in the system has dissipated. A loose inner liner on the lower radiator hose creates an overheating problem that is nearly impossible to diagnose without removing the hose for inspection. Change all hoses every three years or 4000 operating hours. Close heater-hose valves during summer months to reduce the potential for leaks.

Most mechanics prefer to use stainless-steel clamps with a worm gear that engages serrations on the ribbon. It is doubtful whether these clamps provide more security than OEM types. Whatever style of clamp is used, avoid over-tightening. The clamp should compress the hose sheathing and not cut into it.

Accessories

As mentioned previously, cooling systems may incorporate turbo aftercoolers, oil coolers, and various other heat exchangers. Design variations make generalizations difficult. Most of these devices, such as the seawater aftercooler shown in Figure 12-19, can be disassembled for cleaning and, with the proper fixtures, pressurized to detect leaks.

12-19 Yanmar aftercooler uses seawater as the medium of heat transfer.

13
CHAPTER

Greener diesels

The Kyoto protocol, signed in 1997, committed the European Union (EU) to reduce global warming gases by 8% within 15 years. The EU responded by tightening emissions standards on industry and by embarking upon an ambitious program to convert the passenger car fleet to diesel power. Taxes on diesel fuel were reduced and, in a move that would generate much criticism, exhaust emission limits on diesel cars were relaxed.

The results were dramatic. Before things fell apart, diesel cars would capture almost 50% of the Western European market and were poised to do something of the same in the United States. And diesel technology underwent its most intensive development in a hundred years.

Diesel emissions

Carbon dioxide (CO_2) is the premier hot-house gas that, in conjunction with other warming gases, has increased the temperature of the Earth by $0.8°C$ ($1.4°F$) since 1880 (Figure 13-1).

Atmospheric concentrations of CO_2 now stand at 397 parts per million (ppm) up from 372 ppm as recently as 2002. The political consensus is to limit warming to $2°C$, but many climatologists believe it's too late to prevent that outcome. A more realistic target would be $4°C$. That temperature will create massive dislocations in agriculture, facilitate intrusions of tropical diseases into temperate zones, and cause extensive coastal and river flooding. Millions of hungry and thirsty people will be displaced.

CO_2 emissions are a given that no available technology can mitigate. Burn a carbon-based fuel and a predetermined amount of CO_2 is released. Here, diesels have an advantage. Because of their superior fuel efficiency, diesels release 15% to 18% less CO_2 than equivalent SI engines. This was the justification for dieselizing the European passenger-car fleet.

13-1 Global warming is real and accelerating.

Carbon monoxide (CO)   When there is insufficient oxygen in the cylinder, the carbon content of the fuel converts to CO rather than CO_2. Because diesel engines operate with surplus air, relatively little CO is produced except during cold starts and under abrupt acceleration. This is fortunate, because CO is a highly toxic compound. When exposed to a concentration of 1.3% of CO in air, humans die within two minutes.

Hydrocarbons   HC (also known as volatile organic compounds or VOCs) consist of raw or partially burnt fuel and lubricating oil. Some of these compounds are carcinogenic, but in-cylinder controls keep them within acceptable levels.

Diesel emissions would be tolerable if all we have to worry about are CO_2, CO, and HC. Unfortunately, we also must contend with particulate matter (PM) and oxides of nitrogen (NO_X).

Particulate matter   PM, sometimes called diesel particulate matter (DPM), consists of solids, made up carbon particles (soot), sulfates and ash.[1] These particles are normally invisible, but can be seen as black smoke in the exhaust of older or out-of-tune engines (Figure 13-2).

PM with diameters of 1 μm or less (1 μm = 0.000039 in.) lodge deep in the lungs where they increase the risk of lung cancer and cardiovascular disease. At least one clinical study demonstrates a connection between chronic PM exposure and sudden death syndrome. Combustion heat, especially when combined with higher pressures and longer flame duration, is the best way of controlling PM. Particles that escape combustion can be collected in an exhaust filter. However, the smaller nanoparticles elude filtration. The World Health Organization estimates that PM exposure in the EU reduces average life expectancy by 8.6 months.

---

[1]Bugarski, A., et al., "Diesel aerosols and gases in underground mines: Guide to exposure assessment and control," NIOSH Office of Mine Safety and Health Research.

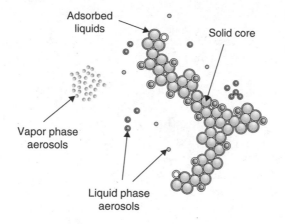

Adsorbed liquids

Solid core

Vapor phase aerosols

Liquid phase aerosols

Elemental carbon core	
Adsorbed hydrocarbons	
Liquid condensed hydrocarbon aerosols	
Sulfates	

13-2 PM consists of organic carbon particles with a coating of carcinogenic aerosols. NIOSH Office of Mine Safety and Health Research.

Oxides of nitrogen NO_X is a blanket term for seven oxides of nitrogen, with NO and NO_2 being the most common. While health professionals consider NO_X to be less of a threat than PM, nitrogen dioxide (NO_2), exposure irritates the airways and has been shown to inhibit lung function in children. There is also a positive correlation between emergency room visits and NO_X exposure and less-than-conclusive evidence that NO_X is a lung cancer precursor. (The greater incidence of lung cancer among affected populations may be associated with PM.) But there is no disputing that NO_X reacts with HC and sunlight to generate smog. The acid rain from these nitrates damages eco systems and adversely affects food crop production.

Oxygen and nitrogen require heat and pressure to combine. To reduce NO_X emissions, it is necessary to quench combustion heat, lower cylinder pressures, mix incoming air with cooled exhaust gases and, in short, make combustion less efficient. And that is precisely what must not be done if PM is to be controlled. What inhibits one pollutant, promotes the other.

Emissions controls

To achieve their goal of making the diesel engine the standard for personal transport, European auto manufacturers were leaders in high-tech emissions controls. Initially efforts involved direct injection, camshaft-driven unit injectors, dished pistons, multiple intake and exhaust valves, delayed injection, and a host of other in-cylinder modifications. When in-cylinder measures proved inadequate, the emphasis shifted to exhaust aftertreatment.

Turbocharging

Turbochargers—beautifully simple machines that in racing applications spin at 300,000 rpm—were described in some detail in Chap. 9. In the context of controlling emissions, the power boost created by a turbocharger enables the engine to be downsized for less weight, reduced friction and, when operated below peak power, for reduced fuel consumption and CO_2 emissions. The power potential is staggering. Back in 1986 when Gran Prix rules permitted supercharging, factory works teams qualified with "kamikaze" 1.5L engines boosted to more than five times atmospheric pressure. Power outputs could reach 1300 hp.

The addition of a turbocharger also permits lower compression ratios, which means less combustion heat and fewer NO_X emissions during periods of moderate loads when the turbocharger idles. Of course, when the turbo comes on line, effective compression increases with a corresponding increase in NO_X. But any decrease in the signature emission of diesel engines is welcome. Variable geometry turbochargers (VGTs) are also the means of exhaust gas recirculation (EGR) delivery, as discussed below.

Much effort has gone into electrification of the turbocharger function. Supplementing exhaust drive with a motor yields low-speed torque and practically eliminates turbo lag. The usual approach is to couple a permanent-magnet motor directly to the turbo shaft. While the advantage is compactness, the motor must be insulated from heat.

What appears to be a more viable alternative for vehicle applications is the use of an electrically driven air compressor in series with the turbocharger. Borg Warner is preparing to put its compressor, known as the eBooster, into series production. The eBooster has a brushless DC motor fitted with rare-metal (samarium-cobalt) magnets to reduce current draw and heat generation. The 12V version draws 200A, the 48V version 130A. Lots of energy is required to compress air.

The eBooster comes on stream during periods of low-speed/load. Once sufficient exhaust energy is available, the turbocharger operates unassisted. The battery could theoretically be recharged by driving the generator with the surplus boost created by the turbocharger at high engine speeds. The generator would replace the wastegate and could also act as an engine brake.

NO_X controls

Initially NO_X could be controlled in-cylinder. But as European and U.S. emissions limits tightened, it became necessary to employ more aggressive techniques.

Exhaust gas recirculation

Recycling inert exhaust gas back into the cylinders quenches the flame below the temperatures where most NO_X forms. Combustion proceeds more slowly and fewer oxygen molecules are available to form NO_X. EGR rates can be as high as 50% at idle and under light loads conditions.

Some stationary and marine-engines introduce EGR into cylinder by leaving the intake valve open during part of the exhaust stroke or by cracking the exhaust valve open during the intake stroke. Light- and medium-duty engines route exhaust gases into the cylinders by way of the intake manifold. Because manifold pressure is only marginally less than atmospheric pressure, these engines use variable geometry turbochrgers (VGTs) to build the necessary backpressure in the exhaust stream. Pressurized exhaust gas is also used to regenerate diesel PM filters and as an engine brake.

Where exhaust gases are sourced for recirculation varies:

- High-pressure systems draw EGR upstream of the VGT, send it through a cooler, and reintroduce the exhaust gas downstream of the compressor (Figure 13-3). Cooling the exhaust makes the air/fuel mixture denser, which raises the effective compression ratio to burn off soot. Most light-duty engines use this approach.
- Low-pressure systems draw exhaust gas from a point downsteam of the filter and cat converter and route it through the compressor (Figure 13-4). Because the exhaust is thoroughly mixed with air in the compressor, less of it must be recycled for the same amount of NO_X reduction. And because the exhaust is filtered, fewer abrasive particles enter the engine. Examples of low-pressure EGR are relatively rare. Mitsubishi Heavy Industries has

13-3 High-pressure EGR systems draw exhaust gas at a point upstream of the turbo, send the gas through a heat exchanger, and into the intake manfold where it mixes with incoming air. This drawing does not include the multiple pressure, flow, and temperature sensors used in actual high-pressure EGR systems.

13-4 Low-pressure EGR systems draw scrubbed exhaust gases from the tailpipe and use the compressor to mix exhaust with air.

experimentally adapted low-pressure EGR to a large marine engine and Volkswagen employs both high- and low-pressure EGR on its TDI and Audi engines.

Pumping corrosive and particle-ridden exhaust gases back into the cylinders is not anyone's idea of good engineering. But if diesel engines are to continue to be sold, they must conform to current European and American NO_X limits, even at the cost of performance and durability. Surfaces exposed to exhaust gases collect soot and lacquer. Exhaust coolers (Figures 13-5 and 13-6) lose efficiency, sensors go dormant, EGR metering valves stick, and intake ports narrow. Some owners remove or plug the EGR system. Aside from legal liabilities, the increased combustion heat and pressure due to the loss of EGR can make what was once reliable transportation exercise in parts acquisition.

Selective catalytic reduction

Selective catalytic reduction (SCR) is a trump card in the struggle against NO_X (Figure 13-7). The technology works by spraying ammonia into the exhaust stream, which then reacts with in the presence of a catalyst to convert NO_X into nitrogen and water. Dry (anhydrous) ammonia can be injected directly into the exhaust or mixed with steam for stationary applications. For reasons of safety, vehicle applications use a mixture of urea and de-ionized water, known as diesel exhaust fluid (DEF) or AdBlue, as the reagent. Urea—CH_4N_2O—breaks down

13-5 A Deere engine with liquid-cooled EGR.

into ammonia (NH_3) in the converter. The longest lived SCRs use platinum or another precious metal as the catalysts, although metal oxides and crystal zeolites can also be used. NO_X reduction takes place at temperatures of 350° to 450°C. For the process to work during cold starts or during long periods of idle, some SCRs have provision for electrical heating. Under optimum conditions, selective catalytic conversion cuts NO_X emissions by more than 90%.

In basic form, the reaction is:

$$NH_3 + NO_X = N_2 + H_2O$$

Stated more precisely to account for the NO and NO_2 components of NO_X, the reaction is:

$$4\,NO + 4\,NH_3 + O_2 = 4\,N_2 + 6\,H_2O + 4\,NH_3 + O_2 = 3\,N_3 + 6\,H_2O$$

13-6 Soot-clogged 6.7L Ford Power Stroke EGR coolers usually throw a P0401 diagnostic trouble code (DTC) that does not clear when other exhaust-related components test normal. Little Power Shop, Bradford, PA.

13-7 Because an SCR is such an effective NO_X scrubber, engines can be tuned for better performance and fuel economy. MECA.

Some reduction of HC and CO also occurs:

$$HC + CO + O_2 = CO_2 + H_2O$$

The delivery system for vehicles consists of:

- A DEF tank large enough to avoid frequent refills. Most systems consume DEF at about 5% of the rate of fuel consumption. The tank assembly includes a heater and fluid level sensors.
- A dosing pump to deliver atomized DEF to the mixer section of the converter.
- A dosing module to match DEF injection with engine speed, load, and exhaust temperature.
- Sensors to monitor SCR conversion rates.
- Dashboard lamps to signal when the DEF tank is low and a means of preventing the engine from starting if the DEF tank runs dry.

The Environmental Protection Agency (EPA) mandates that SCR systems impose no maintenance for 100,000 miles on automobiles and 290,000 miles on heavy-duty trucks.

PM controls

The American Cancer Society suggests that mechanics and farmers take measures to exhaust to protect their lungs from particulate matter. Arranging for good ventilation and standing well clear of idling engines helps. Note that the cloth or surgical masks worn by residents of heavily polluted Asiatic cities offer little protection against the 2.5 micron and smaller particles that pose the greatest health risk. A P95 mask, such as the 3M 8577 particulate respirator, is a far better choice. These disposable respirators are claimed to have 95% filtration efficiency and cost around $30 for a box of 10.

Diesel particulate filter

While low-sulphur fuel and measures to keep crankcase oil out combustion chambers reduce PM formation, a catalyzed diesel particulate filter (DPF) can remove more than 85% of the particles that escape combustion (Figure 13-8). DPFs, used on American pickups and over-highway trucks since 2007 and two years later on European cars, consist of honeycomb Cordierite and silicon carbide elements.

DEFs require periodic regeneration to burn off accumulated soot and reduce exhaust backpressure back to normal levels. There are two ways to do this:

- Passive DPFs clean themselves with exhaust temperatures that must reach 210°C for 40% of engine operating time. These systems are found on construction equipment—Ford, Dodge, and earlier General Motors (GM) pickups, and on some European passenger cars.

Inlet
section

Filter
section

Outlet
section

13-8 DPFs are serious technology. Replacement filter elements sell for around $2000, and costs triple when retrofitting a DPF to an older engine. However, U.S. federal and some state governments underwrite a portion of the retrofit costs.

- Active DPFs regenerate by periodically dousing the filter medium with raw fuel when exhaust backpressure becomes excessive. Chevrolet Duramax systems sense backpressure as the difference in readings between a pressure sensor on the intake side of the DPF and a sensor on the exhaust side. Other manufacturers approximate backpressure as a function of fuel consumption, mileage, or operating hours. This approach does not account for the rapid soot buildup during idle or passive regeneration that results when the engine labors under heavy loads.

   Light-duty truck and passenger car engines regenerate with fuel introduced during the exhaust stroke. Since the fuel ignites well upstream of the DPF, much of the heat dissipates before reaching the filter. In addition, raw fuel contaminates the lube oil, resulting in 5000-mile oil changes. Current Duramax pickups and heavy-duty engines such as the Caterpillar ACERT employ a dedicated injector that delivers fuel to a burner just upstream of the DPF for a claimed fuel saving of nearly 40%. Figure 13-9 illustrates an active DPF used on underground mining equipment.

   Forced regeneration   A DPF must be sent out for cleaning when on-engine regeneration fails to reduce backpressure to normal levels or when an exhaust opacity test indicates excessive levels of PM (Figure 13-10). Most cleaning services remove ash accumulations with a combination of heat, vacuum, and high-pressure air. A more intensive process involves baking the DPF element for eight hours or more in an environmentally sealed oven. Ash deposits are then collected in sealed containers for safe disposal.

13-9 Active DPFs regenerate with the heat provided by an afterburner. Courtesy NIOSH Office of Mine Safety and Health Research.

WARNING: Do not attempt to blow out these filters with shop air. Ash deposits are hazmat.

The rate of ash buildup varies with the application. Passive DPFs clog more frequently than the active types. Worn, oil-burning engines can generate enough ash to clog the filter within hours of cleaning. Yet highway trucks have gone a quarter-million miles between forced regenerations.

13-10 Clean (left) and ash-clogged (right) DPF filter elements. California Air Resources Board.

Diesel oxidation catalysts

Diesel oxidation catalysts (DOCs) oxidize emissions in the presence of an aluminum oxide, palladium, or platinum catalyst. The exhaust stream contains surplus oxygen that converts CO, HC, and the soluble organic fraction (SOF) that adheres to PM into CO_2 and water. SOF reduction has important health benefits, but their elimination merely reduces the mass of particulates by 25% to 40%, and not their number. Even so, these converters eliminate diesel odor and most exhaust smoke.

Since 1995, oxidation catalysts have become almost standard on diesel engines sold in this country. Hundreds of thousands of DOCs have been retrofitted to older engines as their only means of pollution control. The critical requirement is that exhaust gas temperatures must remain above 150°C. The oxidation catalyst usually mounts upstream of the DPF (often in the same canister) to reduce the HC that would interfere with PM filtration, and may be integrated with the muffler or resonator (Figure 13-9). State and federal programs can reduce the costs of retrofits.

Diesel debacle

The EU's efforts met with success in that diesel engines now power half of European passenger-car/light truck fleet. But promised decline of CO_2 emissions did not occur. As Michael Carnes and Eckard Helmers point out:

> … [W]hile petrol-fueled cars emitted continuously less CO_2 over time, diesel cars in Europe stalled emission intensity improvements between 2000 and 2005. New diesel passenger cars in Europe therefore lost most of their CO_2 emission advantage over gasoline cars … Since 2009, diesel car CO_2 emission advantage over petrol cars in Europe became marginal and went down to 1.5% in 2010… CO_2 emissions of new diesel cars registered in Germany had increased since 2001 while at the same time gasoline cars lowered their CO_2 emission intensity.[2]

CO_2 emissions increased because diesel cars became more powerful and fuel hungry. (CO_2 and fuel consumption are inexorably linked.) EURO test cycle protocols made concessions to manufacturers that masked the increase in fuel consumption. Cars were tested after laboratory tuning, with rear seats removed, and body seams sealed with masking tape. As a result, diesel cars give, on the average, 41% less real-world mileage than the sticker indicated. The lack of realism was also evident in the way EURO emissions regulations permitted diesel cars to emit as much as three times more NO_X than their gasoline equivalents.

The 2014 report by researchers at West Virginia University's International Council on Clean Transportation (ICCT) sealed the fate of diesel automobiles. ICCT was contracted by the Council on Clean Transportation, a non-profit research group,

---

[2]Carnes, M. and Eckard Helmers, "Critical evaluation of the European diesel car boom—global comparison, environmental effects and various national strategies" https://enveurope.springeropen.com/articles/10.1186/2190-4715-25-15.

to make on-road emissions tests of a VW Jetta and a Passat, and a BMW X5. The rental cars were fitted with portable emissions test equipment and driven in 16-hour shifts through West Coast urban centers. Exhaust emissions from the BMW were within EPA Tier 2 limits. But that was not true for the Volkswagens. NO_X emissions peaked at 20 times the legal limit for the Passat and 35 times the limit for the Jetta. Yet these cars appeared perfectly legal when tested on a chassis dynamometer by the California Air Resources Board.

In late 2015 the EPA charged the Volkswagen Group with violating U.S. emission laws by installing defeat devices—software designed to cheat emissions tests—on 600,000 VW, Porches, and Audis sold in the United States. All together 11 million of the doctored cars were sold worldwide. VW first denied the allegations, then blamed a cabal of rogue engineers. As evidence and lawsuits mounted, the United States issued criminal indictments. One VW executive, who was foolish enough to leave Germany, was arrested in Florida. Denied bail as a flight risk, he remains in a Detroit jail awaiting trial.

The company eventually pled guilty to conspiracy to violate the Clean Air Act, customs violations, and obstruction of justice charges, and agreed to pay $4.3 billion in fines and penalties and $20 billion to settle lawsuits by U.S. owners of their vehicles. Legal actions in other countries are proceeding.

Researchers at the Massachusetts Institute of Technology calculate that the VW fiasco will cause 1200 premature deaths in Europe where most of the cars were sold.[3]

Shortly after the EPA filed on Volkswagen the agency notified Fiat Chrysler Automobiles that defeat devices had been installed on 2014–2016 3.0L diesel Jeep Grand Cherokees and Dodge Ram 1500s. Fiat Chrysler CEO Sergio Marchionne emphatically disagreed that the devices were illegal. But the damage was done. In an interview with *Auto Express*, Marchionne said, "There are very few things that are certain in the market—apart from one, and that is that small-displacement diesels are dead."[4] And indeed they are. Mass-market diesels lack the profit margin to support clean exhausts. Subsequently, other diesel makers including BMW, Mercedes Benz, Peugeot and Alfa-Romeo have come under investigation.

Disenchantment is confined to light-duty, consumer-level diesels. Medium- and heavy-duty engines will be the norm for the foreseeable future. No other power source can match their fuel economy, torque output, and durability (Figure 13-11).

Biodiesel

Biodiesel consists vegetable or animal fat converted to acid methyl esters (FAME) and should not be confused with straight vegetable oil (SVO also known as pure plant oil, or PPO). FAME has less viscosity and a lower boiling point than SVO and can be blended with petroleum-based diesel. B5 or B20, consisting of 5% or 20% FAME, are

---

[3]Massachusetts Institute of Technology, "Volkswagen's excess emissions will lead to 1,200 premature deaths in Europe." *ScienceDaily*, March 10, 2017.

[4]Jacobs, C. "FCA CEO Sergio Marchionne plans on replacing diesel fiats with hybrid models," *Auto Express*, March 10, 2017.

13-11 A WinGD 10X92 marine engine under construction in South Korea demonstrates why diesel engines will remain as prime movers. The 12-cylinder version of this engine develops 62,840 kW at 80 rpm with a specific fuel consumption of as little as 158 grams/kW-hour. The tiny human figures on the left provide a sense of scale. (© WinGD)

the most common diesel blends sold in the United States. Benefits include reduced CO, HC, and PM emissions (Figure 13-12. The FAME component releases no net CO_2. That is, CO_2 resulting from combustion will be absorbed by the next crop of feed stock.

Contrary to some reports, the Argonne National Laboratory has established that FAME has a positive energy balance. For each unit of energy absorbed during the production of the vegetable base, 4.56 units are released during combustion.

All diesel engines can safely use B5, and most modern engines can have been adapted to run on B20. Notable exceptions are recent Mercedes-Benz engines, rated only for B5. Deutz AG certifies its agricultural engines for B100. Table 13-1 lists potential problems.

FAME is hard on fuel filters. The fuel dissolves varnish deposits, oxidizes rapidly, and attracts water and bacterial growth. "Soapy" deposits on common-rail injectors have become common since the introduction of ultra-low sulphur fuel (ULSF). But the role FAME plays in the formation of these deposits has not been established.[5] A comparison study of Ford and Mack diesel trucks operated on neat diesel and B20 for 4 years showed no significant differences in engine wear.[6]

---

[5]"Internal diesel injector deposits," CRC Report No. 665, Oct., 2013.

[6]Fraer, R., et al., "Operating Experience and Teardown Analysis for Engines Operated on Biodiesel Blends (B20)," SAE 2005-01-3641.

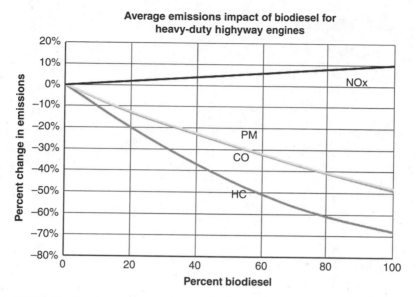

13-12 Biodiesel emissions impacts.

Table 13-1. Possible effects of FAME on fuel systems

FAME characteristics	Failure mode	Effects
Fatty acid methyl esters in general	Swelling, softening, and cracking of elastomers, especially nitrile rubbers	Fuel leaks Filter clogging
Free methanol	Aluminum and zinc corrosion	Fuel system corrosion
	Low flash point	Safety concerns
Free water	Corrosion	Fuel system corrosion
	Reversion of FAME to fatty acids and methanol	Filter clogging
	Bacterial growth	
Free methanol	Corrosion of nonferrous parts	Fuel system corrosion
Free glycerin	Sediments	Filter clogging
	Lacquer buildup	Injector coking
High viscosity at low temperatures	Excessive heat in rotary distributor pumps	Premature failure of pumps and related parts
	Higher stresses on pump-drive systems	Poor nozzle atomization
Solids	Reduced lubrication	Premature fuel system failure

Straight vegetable oil

"The use of vegetable oils for engine fuels may seem insignificant today, but such oils may become in the course of time as important as petroleum and the coal tar products of the present time."

– Rudolf Diesel

One of the attractions of the 1900 Paris World Fair was a diesel engine running on peanut oil. Rudolf Diesel hoped that his engines would provide Europe with energy independence. After it became clear that straight vegetable oil could not compete on price with petroleum, the inventor struggled for years to adapt his engines to run on coal dust.

Rapid increases in oil prices during the 1970s and the burgeoning environmental movement sparked a new interest in straight vegetable oils. A leading figure in the movement was Ludvid Elsbett (1913–2003), an engineer of remarkable creativity who held more than 400 patents. In 1937 Elsbett was hired by Junkers Motorenwerke AG to assist in the development of the gasoline direct injection system used on the Jumo 211 aircraft engine. As a point of interest, the fuel system consisted of 1575 individual parts, some of which were lapped to light-wave-length precision.[7]

Subsequently, Elsbett and his sons opened a shop in Bavaria that became the first commercial source of SVO auto engines. Hundreds of conversions were made, ranging from all-mechanical indirect injection engines to VW common-rail TDIs and at least one locomotive. To achieve OEM levels of driveability and reliability, the firm made comprehensive modifications to the engines, modifications that included special injectors, bi-metallic pistons, and oil cooling. Because oil has a higher boiling temperature than water, oil-cooled engines run hotter and lose less heat to the coolant.

SVO poses problems for less comprehensive conversions. One of the most intractable problems is the way vegetable oil carbons over injectors and combustion chambers. Figure 13-13 graphs the coking effects of sunflower oil against No. 2 Diesel. And even with frequent de-carbonization skewed nozzle spray patterns and contaminated lube oil pose risks of reduced engine life.

SVO congeals into a thick syrup at low temperatures. As shown in Figure 13-14, sunflower oil is an order of magnitude more viscous than No. 2 Diesel at 18°C. SVO also attacks seals, hoses, and other elastomer parts.

Use of the fuel peaked near the end of the last century before diesel engines became so complex. Rape-seed oil was the favored feedstock in Europe, sunflower-seed oil in this country. Some enthusiasts went further and ran their cars on waste cooking oil, unprocessed except for filtration.

Those individuals who continue to use these fuels reduce its viscosity with heat or by thinning with No. 2 Diesel, alcohol, or even gasoline. Older IDI engines with Bosch inline pumps and pintle injectors adapt most easily. Their high-pressure pumps are nearly indestructible and, when necessary, the injectors can be easily disassembled for

---

[7]Green, P.M., "A penny for your thoughts, Miss Shilling," *Aeroplane Monthly*, February, 1997, p. 39.

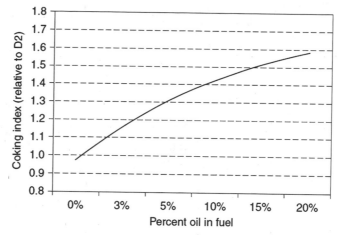

13-13 How rapidly engine parts carbon over increases with the percentage of SVO added to conventional fuel.

cleaning. CAV, Delphi, Lucas, Stanadyne, or RotoDiesel distributor-type pumps should be avoided. Most SVO adaptations use diesel fuel for starting and have a second tank heated by engine coolant for the vegetable oil. A thermostatically controlled resistance heater installed at the suction side of the injector pump can take some of the strain off the pump and its drive mechanism. Manually controlled glow plugs that remain energized until engine temperatures stabilize reduce the battery draw during cold starts.

Even so there is a role for SVO. In the summer of 2006, a group of Brazilian university students and technicians arrived at Vila Soledade, a rural community on

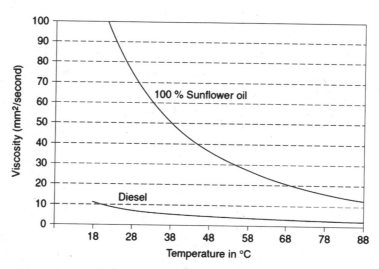

13-14 At room temperature, sunflower oil exhibits ten times the viscosity of No. 2 Diesel.

1- Diesel engine
2- Generator
3- Palm oil service tank
4- Connection to the diesel oil system
5- 3-way valve
6- Preheater

7- Transference pump
8- Palm oil filter
9- Heating tank
10- Electric heater
11- Manual pump
12- Engine fuel
13- Injection pump

13-15 The SVO kit developed by the Brazilian government.

the banks of the Amazon.[8] Vila Soledade is remote from the national grid and must generate its own electrical power. In theory, the local gen-set could provide the 700 townspeople with electricity for five or six hours a day. But diesel fuel is expensive for farmers living barely above the subsistence level and their antique gen-set frequently broke down. They would wait weeks for replacement parts.

The student team brought with them a Brazilian-built MWD TD229EC-6 generator powered by six-cylinder, turbocharged engine, modified to run on unprocessed palm oil. This oil is local product that costs nothing except the labor involved in its extraction.

As shown in Figure 13-15, the generator engine circulates coolant through a holding tank to preheat the oil to 65°C (140°F). A transfer pump moves the warm oil through a filter to a second tank where it is heated to 85°C (185°F) for injection. With heat, the palm oil moves from the viscous esterarine phase to the more pumpable oleine phase. Sensors report fuel pressure and ambient, fuel and exhaust temperatures, which are recorded daily by the operator.

Vila Soledade now has electricity for six hours every day. To prolong filter life the engine is run for a half hour on diesel fuel immediately after starting and just before shutdown. The rest of the time it operates on raw palm oil.

Stack emissions on diesel fuel and palm oil were found to be almost identical, except that vegetable oil produced no sulfur oxide. An analysis of the lube oil allowed

---

[8]Coello, S. et al. "Energy from vegetable oil in diesel generator—results of a test unit at Amazon region," Brazilian Reference Center on Biomass, COMEC.

change periods to be extended to 200 hours, which was surprising, since straight vegetable oil has a reputation for contaminating lube oil. Misfiring developed after 600 hours due to dirty injectors. The injectors were changed and at 800 hours the head was lifted to decarbonize the combustion chambers. The major problem was fuel filters that gummed over and required frequent changes.

Reliable and affordable electricity has transformed the community. With the money saved on fuel, the villagers purchased equipment to process acaí, one of the major crops of the region, and many families have purchased lamps, television sets, refrigerators, and freezers. Most adults now attend night school.

Natural gas

Natural gas (NG) is the most widely used surrogate for diesel fuel. Advantages include:

- Lower emissions—because NG contains less carbon than diesel fuel, the microscopic spheres of soot that make up PM are reduced by as much as 70%. And because NG displaces oxygen in the combustion chamber, it burns cooler than diesel fuel to cut NO_X emissions by half. Control of HC and CO usually requires exhaust aftertreatment.
- Predictable conversion costs—bolt-on kits adapt light- and medium-duty compression ignition (CI) engines to NG while retaining the ability to run on diesel fuel. Large marine and industrial engines can be set up to burn both fuels or NG only. Dedicated NG conversions require spark ignition and low-compression pistons.
- Lower fuel cost—in general NG costs less than conventional diesel fuel. However, markets fluctuate and the price advantage is not always present.

Compressed natural gas (CNG)

Most of NG-powered vehicles run on compressed natural gas stored at between 2400 and 3600 psi in cylindrical pressure vessels (Figure 13-16). CNG cylinders come in four configurations:

- Type 1—all-steel construction without a composite reinforcement. Similar to welding-gas cylinders, these tanks are the least expensive and heaviest.
- Type 2—aluminum or steel cylinder with a fiberglass, carbon or armid wrapping over the tubular section. The metal tanks ends are uncovered. Type 2 tanks are 25% lighter than Type 1 steel tanks.
- Type 3—the aluminum or steel liner is fully encased with the same composite wrapping used on Type 2 cylinders. Type 3 cylinders have the best tradeoff between weight and cost.
- Type 4—similar to Type 3 except that the liner is made of a polymer.

Rooftop CNG fuel system
Allows for larger-capacity system but presents overall
vehicle height and center-of-gravity challenges

Back of cab CNG fuel system
Allows for larger-capacity system but
may increase wheelbase or overall
length of vehicle

Frame-mounted CNG, LNG, or propane fuel system
Vehicle wheelbase can be a factor in system capacity

13-16 Fuel storage options. Fuel tanks are subject to what firefighters call BLEVES (boiling
liquid expanding vapor explosions). External mounts enable the tanks to be cooled with fire-
hoses and make it less likely that leaking gas can collect in the vehicle bodywork.

The range limitations of CNG and the paucity of refueling stations limit this
technology to vehicles that circulate around a central hub. School buses and trash
pickup trucks are prime candidates, in no small part because of the federal and state
grants to municipalities.

Fast-fill stations serve retail customers who expect rapid refueling, made possible
drawing down CNG stored onsite in pressure vessels (Figure 13-17). A 20-gallon

Fast-fill station

13-17 Fast-fill CNG stations satisfy the need of retail customers for rapid refueling.

Time-fill station

13-18 Time-fill CNG stations are appropriate for fleet operations.

equivalent of diesel fuel can be discharged within minutes. But the heat expansion during rapid fillups reduces cylinder capacity.

Time-fill stations provide in-house refueling for fleets that are idled for extended periods (Figure 13-18). Unlike fast-fill stations, the compressor pumps CNG directly into vehicle tanks, an operation that can require several hours. The buffer storage vessel shown in the drawing contains a small amount of CNG for topping off tanks without the costs associated with starting the compressor. Because compression proceeds incrementally, the gas absorbs little compression heat, which results in a denser charge and a greater vehicle range for the same work of compression. In addition, refueling can be scheduled to take advantage of off-peak electricity billing.

LNG

Super-cooling natural gas to −260°F converts it into a liquid with more than twice the energy density of 3600-psi CNG. Heavier hydrocarbons are stripped off during processing to yield a product that consists of nearly pure methane.

Unlike CNG, which has an infrastructure in the form of pipelines, LNG must be trucked to refueling stations (Figure 13-19). Because of higher transportation and processing costs, LNG is more expensive than compressed gas. As of January 2017, the average price of CNG sold in the United States was $2.38 diesel equivalent gallon (DEG). LNG cost $2.53. LNG also requires special handling precautions. Personnel at risk of contact with the cryogenic liquid need protective clothing and face shields.

LNG is stored at pressures between 60 and 230 psi in double-walled, vacuum-insulated tanks, not unlike giant thermos bottles. But no insulation is perfect and some heat gets through the vacuum barrier. LNG boils at −259°F, or 1 degree more than its storage temperature. When the pressure of the boiled off gas approaches the design limit of the tank, a relief valve opens to vent the vapor into the atmosphere.

13-19 LNG refueling stations must be supplied by truck.

Of course, venting should be avoided since it wastes fuel and accelerates climate change. Methane has 25 times more global-warming effect than CO_2.

The fuel consumed by a running engine lowers tank pressure faster than boil-off increases the pressure. Stop the engine, and pressure slowly builds. The number of days before which the tank vents is called the "hold time." That specification can be misleading because it is based on a full tank of fuel. Heat enters at a constant rate, independent of the fuel level. If the tank is partially full, there is less fuel to absorb the heat and gas pressure increase more rapidly than indicated by the hold time specification. (You can boil a half-cup of water faster than a full cup.) As a rule of thumb, LNG has a storage life of 1 week. Operators can get a more accurate idea of hold time if the instrumentation reports the fuel level, tank temperature and tank pressure with the engine shut off.

With one quarter of the bulk and a fraction of the weight of CNG systems, LNG is the choice for Class 7 and 8 over-the-highway trucks and for marine applications. Three-hundred DEG gallons of LNG gives a tractor trailer truck the range of 600 miles.

Dual-fuel engines

Dual-fuel engines run either on 100% diesel fuel or on a combination of NG and small amounts of diesel. When running on NG, most dual-fuel systems pass a metered amount of gas into the intake manifold. These systems are described as low pressure (Figure 13-20). Some large dual-fuel engines introduce gas directly into the cylinder by means of a dedicated injector. The point in the compression stroke when gas is admitted varies. Energy Conversions Inc., a major player in dual-fuel railroad conversions, produces a system that injects gas early in the stroke before much compression pressure has built (Figure 13-21. High-pressure systems, like the one used on the MAN BW ME-GI two-stroke marine engine, inject gas late, against the full force of compression.

However gas is admitted, dual-fuel engines initiate combustion with an injection of a small pilot charge of pure diesel. Methane, the major component of natural gas, requires 580°C. (1076°F) to ignite, which is a higher temperature than produced by compression. The pilot charge, which has an ignition temperature of 210°C (410°F), acts as a kind of spark plug.

Diesel fuel burns quickly into controlled explosion. NG burns progressively like a grass fire. The difference in burn rates does not matter for large, 200-rpm marine engines,

Scheme of the converted engine

13-20 Low-pressure systems are relatively inexpensive and are less likely to develop leaks. Should a rupture occur, the liberated gas has less expansion energy. However, power outputs at full load are somewhat less than those obtained when operating on pure diesel. ComAp

some of which burn 99% NG under full load. But small engines of the sort used for motor vehicles must compensate for the slow burn rate of NG by reducing the amount of gas admitted at high speed. Otherwise, the flame front would not keep pace with piston movement and combustion would continue into the exhaust stroke. These engines have a maximum NG substitution rate of about 80%. The substitution rate is also reduced at low speeds when the gas/air mixture verges on the lean limit of combustion.

This technology may represent, at least in the short term, the future of large diesel engines. Marine operators appreciate the ability to meet emissions regulations with clean-burning natural gas and the flexibility to use diesel when LNG is not available. Railroads are another prime candidate. According Burlington Northern Santa Fe railroad CEO Matt Rose, LNG may be "the next big opportunity for taking costs out of operations."[9]

Canadian National Railway has two 3000-hp EMD SD40-2 locomotives with LNG/diesel conversion kits supplied by Energy Conversions Inc. in regular service. GE Transportation is currently testing its NextFuel™ retrofit kit that, with LNG substitution rates of as high as 80%, promises to cut fuel costs by half. Converted locomotives run for twice the haul distance between refuelings, which simplifies the infrastructure.

---

[9]Vantuono, W.C., "Locomotives: Is LNG the next generation?" *Railroad Age*, September 24, 2014.

Boost air cleans the chamber "scavenging"

Gas injected to the cylinder during compression

Gas mixes with air during compression

Ignition is started by a small amount of diesel being injected

Gases expand from combustion, pushing the piston down

Exhaust leaves the chamber through valves when expansion is completed

13-21 Energy Conversions Inc. (Tacoma, WA) was the first to produce a commercially viable dual-fuel locomotive and has since branched out into oilfield, marine and other stationary engines. The illustration shows how an EMC locomotive engine is adapted to burn NG. The conversion kit includes special pistons, cylinder heads, gas injectors, electronic controls and an auxiliary hardware. This technology can reduce CO_2 emissions by 30% and NO_X by 70%.

Renewable natural gas (RNG)

NG comes about as the result of bacterial action on decaying organic matter. We drill for it in ancient sea beds that once pulsed with microscopic and plant life. At present rates of consumption, U.S. reserves of NG should last for another century. These reserves, collected over 550 million years, are non-renewable on a human scale.

However, municipal landfills and sewage pools contain significant amounts of renewable natural gas. The EPA Landfill Methane Outreach Program reports that the U.S. has 654 ongoing methane recovery projects and the potential for 400 more. While RNG cannot meet all our energy needs, it is a significant source of methane that if left untapped, finds its way into the atmosphere where it accelerates global warming.

About six months must pass for organic waste to yield useful amounts of NG and 20 years for production to peak. Figure 13-22 illustrates how landfill gas is collected and used to generate electricity. The 13,000 gal/day of LNG California's Altamont Landfill produces powers the processing plant and 300 trucks that supply the landfill with refuse.

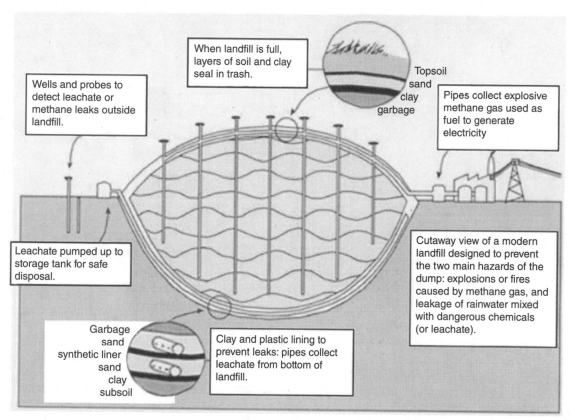

13-22 How methane is harvested from a solid-waste landfill.

Index